Chinese Energy Futures and Their Implications for the United States

Chinese Energy Futures and Their Implications for the United States

George Eberling

LEXINGTON BOOKS
Lanham • Boulder • New York • Toronto • Plymouth, UK

Published by Lexington Books
A wholly owned subsidiary of The Rowman & Littlefield Publishing Group, Inc.
4501 Forbes Boulevard, Suite 200, Lanham, Maryland 20706
www.lexingtonbooks.com

Estover Road, Plymouth PL6 7PY, United Kingdom

British Library Cataloguing in Publication Information Available

Library of Congress Cataloging-in-Publication Data
Eberling, George.
Chinese energy futures and their implications for the United States / George Eberling.
p. cm.
Includes bibliographical references and index.
ISBN 978-0-7391-6568-3 (alk. paper) — ISBN 978-0-7391-6570-6 (electronic)
1. Petroleum industry and trade—China. 2. Energy security—China. 3. China—Foreign economic relations—United States. 4. United States—Foreign economic relations—China. I. Title.
HD9576.C52E235 2011
333.8'2320951—dc23 2011035276

™ The paper used in this publication meets the minimum requirements of American National Standard for Information Sciences—Permanence of Paper for Printed Library Materials, ANSI/NISO Z39.48-1992.

Printed in the United States of America

This book represents the influence of my life of
my mother. She constantly urged me to succeed
by the wisdom of her own example. Without her
encouragement, I would not have stayed on course.

Contents

List of Tables and Figures

TABLES

FIGURES

More information on the study that inspired this book, as well as more of the raw, original data from the study itself, may be found at http://www.princeandpanda.com.

List of Abbreviations

AFRICOM	United States Africa Command
ARF	ASEAN Regional Forum
ASEAN	Association of South East Asian Nations
BAP	Beijing Action Plan
BBL	Barrels per Year
BMENA	Broader Middle East and North Africa
BRIC	Brazil, Russia, India, China
CBO	Congressional Budget Office
CENTO	Central Treaty Organization
CJTF-HOA	Combined Joint Task Force-Horn of Africa
CNA	Center for Naval Analysis
CNG	Compressed Natural Gas
CNPC	China National Petroleum Corporation
CNOOC	China National Offshore Oil Corporation
CRS	Congressional Research Service
CSIS	Center for Strategic and International Studies
CTF	Combined Task Force

CUPET	Cubapetroleo
CVBG	Carrier Battle Group
CVN	Nuclear Aircraft Carrier
DOD	U.S. Department of Defense
DOE	U.S. Department of Energy
DOS	U.S. Department of State
EEZ	Economic Exclusion Zone
EIA	Energy Information Administration
EU	European Union
FDI	Foreign Direct Investment
FOCAC	Forum on China-Africa Cooperation
HEV	Hybrid Electric Vehicle
HoA	Horn of Africa
IEA	International Energy Agency
IMF	International Monetary Fund
IOC	Independent Oil Company
MEND	Movement for the Emancipation of Niger Delta
MES	Market Economic Status
MOOTW	Military Operations Other Than War
NATO	North Atlantic Treaty Organization
NGO	Non Governmental Organization
NOC	National Oil Company
NORAD	North American Aerospace Defense Command
NORINCO	China North Industries Corporation
OAPEC	Organization of Arab Petroleum Exporting Countries
OECD	Organization of Economic Co-operation and Development
OPEC	Organization of Petroleum Exporting Countries

PLA	People's Liberation Army
PLAN	People's Liberation Army Navy
PRINCE	Probe, Interact, Calculate, and Execute
SCO	Shanghai Cooperation Organization
SEATO	Southeast Asia Treaty Organization
SEZ	Special Economic Zone
SINOPEC	China Petroleum and Chemical Corporation
SIPRI	Stockholm International Peace Research Institute
SLOC	Sea Lines of Communication
SSBN	Nuclear Ballistic Missile Submarine
SSGN	Nuclear Guided Missile Submarine
SSN	Nuclear Attack Submarine
SPR	Strategic Petroleum Reserve
TSCTI	Trans-Sahara Counter-Terrorism Initiative
UN	United Nations
UNDP	United Nations Development Program
WMD	Weapons of Mass Destruction

Chapter 1

Introduction

HISTORICAL OVERVIEW OF CHINA'S OIL ENERGY DEPENDENCY

On June 23, 2005, the China National Offshore Oil Corporation (CNOOC), through its Hong Kong subsidiary (CNOOC Ltd.), announced an unsolicited bid to acquire the U.S. energy company Unocal for $18.5 billion in cash. After considerable opposition from members of Congress, CNOOC withdrew its bid, allowing Chevron Corporation to buy and merge Unocal as a subsidiary company.

While CNOOC's intentions may have been honorable, the affair raised important questions regarding U.S. national interests, such as security (protection of life and property), prosperity (protection of economic well-being and commerce), and value preservation (protection and projection of core values of democracy, freedom, human rights, etc.).[1] Further, the incident was what many experts believed to be a microcosm of a larger energy resource competition between China and the United States overshadowed by misperception and having major implications for U.S. global power and influence in light of China's fuel oil energy dependency.

CNOOC's acquisition bid is one of many recent examples in which China was competing to acquire foreign crude oil and petroleum because its existing refinement capacity was not meeting its current oil energy demand. Thus, China was no longer oil energy resource self-sufficient, because China's domestic oil energy consumption had exceeded its domestic oil energy production (see Figures 1.1 and 1.2).

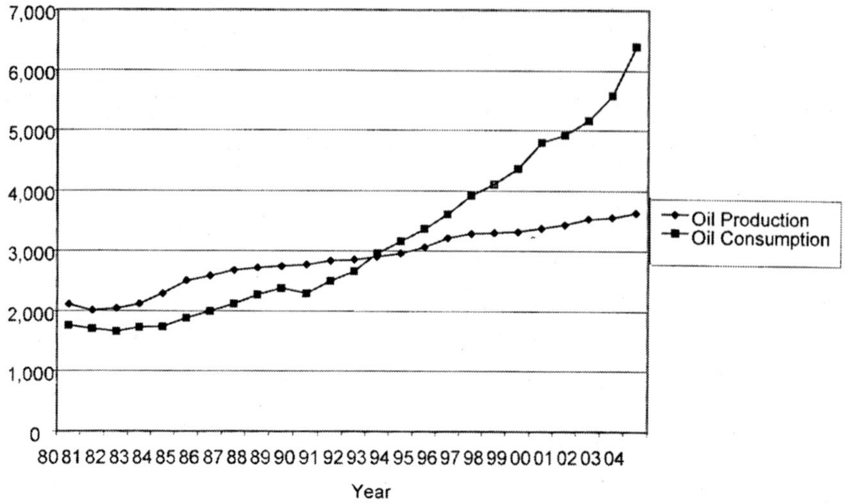

Source: International Energy Agency

Figure 1.1 China's Daily Oil Consumption vs. Production (1980–2004)

THREE OIL ENERGY FUTURES OR SCENARIOS

China's desire to pursue global oil competitively in the wake of its crude oil and petroleum dependency raises some important questions for the United States. Of great importance is the question, what will be the strategic implications of China's oil energy needs on the United States under current conditions—that is, a scenario of competitive dependency? The competitive nature of United States–China energy needs could potentially lead in two directions: competition for scarce oil resources or cooperation for those scarce resources. Further, what are the strategic implications of China's oil energy needs for the United States under other conditions, such as a competitive surplus (nondependency) scenario or a cooperative surplus (nondependency) scenario? In other words, how will each energy future or scenario under the conditions of dependency and nondependency affect the United States globally on a political, economic and military scale?

Other questions to be asked include:

- What events will be associated with each oil energy future or scenario?
- What will be the most likely Chinese oil energy future or scenario?
- What will be the most likely U.S. policy responses?

It is essential that these questions and more be fully addressed and answered because the absence of any definitive conclusions on China's oil energy future

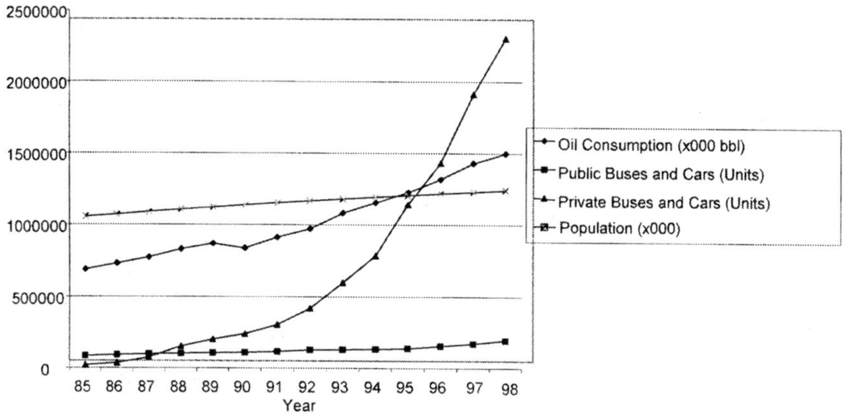

Source: National Bureau of Statistics (China)

Figure 1.2 China's Oil Consumption, Transportation, and Population

under the conditions of dependency and nondependency presents a significant gap in the existing body of knowledge with respect to global energy security trends. This book will hold significant implications for policymakers and government officials alike and will help them to make more informed foreign policy decisions, especially those that influence or affect global energy security. To that end, three scenarios are analyzed to determine which is the most likely policy course between the United States and China as they vie for current and future oil sources:

- A *competitive dependency scenario* is one in which China imports foreign crude oil and petroleum to sustain economic growth, competition exists for global oil supplies between China and the United States (China and the United States import crude oil and petroleum from the same producers), and competition exists between China and Brunei, Malaysia, the Philippines, Taiwan and Vietnam in the South China Sea region.
- A *competitive surplus scenario* is one in which China relies heavily on fuel-saving policies and technologies (hybrid, electric, compressed natural gas) to decrease consumption and increase production and competition exists between China and Brunei, Malaysia, the Philippines, Taiwan and Vietnam with respect to oil exploration, extraction, refinement and transportation in the South China Sea region. China explores for oil in contested territorial waters, and China relies predominantly on domestically produced oil.
- A *cooperative surplus scenario* is one in which China relies heavily on fuel-saving policies and technologies (hybrid, electric, compressed natural gas) to decrease consumption and increase production, and cooperation

exists between China and Brunei, Malaysia, the Philippines, Taiwan and Vietnam with respect to crude oil exploration, extraction, refinement and transportation in the South China Sea region. China explores for oil in resolved territorial waters, and China relies predominantly on domestically produced oil.

In the competitive dependency scenario, China uses a "Go Out" policy or "Going Global" strategy to secure energy resources abroad. This policy seeks to employ China's foreign reserves by acquiring assets overseas and equipping its domestic firms (such as China's national oil companies) and their management with international experience so that they can take the competition to the home markets of foreign nations and compete better in mainland China's own domestic market. CNOOC's $18.5 billion proposed purchase of Unocal is one example of the "Go Out" policy. In the Unocal affair, the Chinese government (a state actor) provided the financial leverage for CNOOC to purchase Unocal. However, the U.S. Congress objected, and CNOOC withdrew its bid to acquire Unocal.

In the nondependency scenarios (competitive surplus and cooperative surplus) China does not use a "Go Global" strategy because China does not import foreign oil. It is not necessary to use a "Go Out" policy because China uses fuel-saving technologies and domestic oil resources.

The scope of this book is limited to three possible Chinese oil energy futures and their strategic implications on the United States using scenario analysis because this perspective paints the broadest picture capturing the full range of potential alternative futures even through there will be permutations and combinations of the three futures.[2] While other energy futures offer considerable merit to address China's dependency of foreign crude oil and refined petroleum such as scenarios based on the use of biofuel technologies (ethanol and/or manure processing) or other "unconventional" fossil fuel resources such as heavy oil, tar sands, oil shale, or gas resources—which are not counted at proven oil reserves[3]—such scenarios would involve exhaustive qualitative and quantitative research and data that would extend well beyond the scope of the investigation. My research leads me to believe that these are very low-probability events and thus futile to explore. Further, such futures would deviate considerably from what this research seeks to demonstrate in the realm of possible oil energy futures in light of estimated levels of global oil production, oil consumption, and proven oil reserves. In this respect, British Petroleum estimates that the world has approximately 41 years of conventional crude oil reserves remaining.[4] This book seeks only to examine and determine those energy futures that have the highest probability of occurrence and the significance for future United States–China relations.

I considered using a cooperative dependency oil energy future for this book, in which China would be dependent on foreign oil to sustain economic growth but would not import oil from the same oil producers as the United States. However, China would not be able to sustain economic growth in this future because it would not be able to meet its annual oil energy consumption demands. China imports over 80 percent (approximately 800 million barrels per year) of its oil from the same countries as does the United States under normal conditions, as depicted in the competitive dependency scenario (See Figure 1.3). The left cylinder represents China's annual oil imports under competitive conditions.

China would import 20 percent (approximately 200 million barrels per year) or less from countries that do not export oil to the United States (see Figure 1.3). The right cylinder represents China's annual oil imports under

Table 1.1 Three Chinese Oil Energy Futures or Scenarios

Competitive Dependency Scenario
Oil production lags behind consumption.
Competition exists for global oil supplies between China and the United States (China and the United States import crude oil and petroleum from the same producers).
China imports foreign crude oil and petroleum to sustain economic growth.
Competition exists between China and Brunei, Malaysia, the Philippines, Taiwan ,and Vietnam with respect to oil exploration, extraction, refinement and transportation in the South China Sea.
Competitive Surplus Scenario (Non-Dependency)
Oil production matches or exceeds consumption.
Competition exists between China and Brunei, Malaysia, the Philippines, Taiwan, and Vietnam with respect to oil exploration, extraction, refinement, and transportation in the South China Sea.
China explores for oil in contested territorial waters; China relies predominantly on domestically produced oil.
China relies heavily on fuel-saving policies and technologies (hybrid, electric, compressed natural gas) to decrease consumption and increase production.
Cooperative Surplus Scenario (Non-Dependency)
Oil production matches or exceeds consumption.
Cooperation exists between China and Brunei, Malaysia, the Philippines, Taiwan, and Vietnam with respect to crude oil exploration, extraction, refinement, and transportation in the South China Sea.
China explores for oil in resolved territorial waters; China relies predominantly on domestically produced oil.
China relies heavily on fuel-saving policies and technologies (hybrid, electric, and compressed natural gas) to decrease consumption and increase production.

cooperative conditions. Because daily oil consumption exceeded daily oil production since 1993 and almost doubled in 2008, China would have to implement draconian public policy measures to sustain such economic growth. Such policies would include but not be limited to decreasing oil consumption and increasing oil production such as petroleum rationing or tapping into its strategic petroleum reserve (SPR). China would have to consider fuel-saving automotive technologies such as hybrid electric vehicles (HEVs), compressed natural gas (CNG) vehicles, or even fuel cell vehicles. Because oil is critical to China's economic growth, it would be unrealistic to expect China not to import oil from the same oil-producing countries as the United States does.

Table 1.2 provides a graphic representation of the dependency oil energy futures (competitive dependency and cooperative dependency) and the non-dependency oil futures (competitive surplus and cooperative surplus). Note that the cooperative dependency oil energy future was not considered.

SUPPORTING ARGUMENTS

Several arguments help to inform our understanding of future Sino-American relations with respect to oil resources. The first argument relates to the anxieties and growing danger of conflict associated with acquisition of energy resources; the second argument relates to "resource nationalism," which might be defined as the management of energy flows in accordance with vital state interests.

In *Blood and Oil: The Dangers and Consequences and America's Growing Dependency on Imported Petroleum; Resource Wars: The New Global*

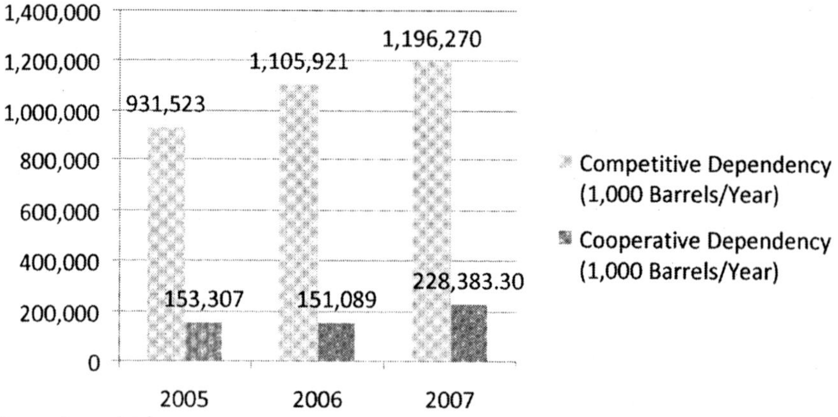

Source: Energy Intel

Figure 1.3 China's Total Annual Oil Imports

Table 1.2 Chinese Oil Energy Futures or Scenarios

Oil Energy Future/Scenario #1: *Competitive Dependency* (Oil Dependency)	Oil Energy Future/Scenario #2: *Competitive Surplus* (Oil Non-Dependency)
Oil Energy Future/Scenario #4: *Cooperative Dependency* (Oil Dependency) *This future/scenario was not considered.*	Oil Energy Future/Scenario #3: *Cooperative Surplus* (Oil Non-Dependency)

Landscape of Global Conflict; and Rising Powers, Shrinking Planet: The New Geopolitics of Energy, Michael T. Klare writes extensively about the problems associated with the acquisition of energy resources by the world's most consuming nations, such as China and United States. Using case studies as evidence, he argues that petroleum production is in a state of decline because the global energy industry will not be able to increase the output of crude oil to satisfy the ever-rising levels of consumption by large industrial nations and that global production capacity arises from a steady decline in the rate of new oil field discovery. Klare contends that competition for access to oil among the world's major consuming nations (such as China and the United States) is expected to become increasingly fierce in light of the fact that the global energy industry will not be able to ramp up production to satisfy the rising demand for oil. In addition, Klare argues that a shift has occurred in the center of gravity of world oil production from the developed world to the developing world, which has profound implications for energy-seeking nations because it entails a heightened risk to the uninterrupted flow of energy supplies.[5] Heightened risk of instability and violence in the oil-producing areas (due to the specific targeting of oil facilities by anti-Western extremists in recent times) will increase the risk of disruptions in the flow of oil from the supplying to consuming nations. Alarmed by this prospect, Klare anticipates that the United States will increasingly rely on military instruments to safeguard the flow of oil, which is known as the "militarization of energy management."[6]

In a similar vein, in the *Alliance Curse* and *Capital and Collusion*, Hilton Root employs case studies to write about dangers associated with asymmetric alliances between First and Third World countries. He argues that throughout history, "an alliance between a greater power and a weaker strategically significant dictator traps the dominant partner into unnecessary conflicts and moral compromises."[7] Although Root does not refer specifically to oil as the source of the weaker nation's strength in the unlikely relationship or even China's asymmetric relations with the developing world in the post-Cold War era, he demonstrates how these types of relationships undermine political reform, economic development, and the rule of law in the weaker nations of developing Asia and Latin America. Essentially, political and social instability results, and economic development is stymied.

The case studies presented by Klare and Root are quite illustrative regarding the anxieties and dangers of First and Third World relationships or alliances with large industrial nations. Klare goes a step further to discuss the specific anxieties and dangers associated with the acquisition of energy resources by world's most consuming nations, such as China and the United States. While Klare's work is intuitive in nature and qualitative in approach, he does not provide empirical evidence to support his central argument. His work would have been better served by quantitative analysis.

In this respect, Table 1.3 shows a direct relationship between the world's top 10 oil producers, their annual oil production, their oil exports to China and to the United States (in thousands of barrels per year) and the level of freedom associated with each oil producer. The figures in bold represent a decrease in the percentage of total annual imports from the previous year (2006), and the figures in black represent an increase the percentage of the total annual imports from the previous year (2006).

Freedom House reports that of the top 10 oil-producing nations for China and the United States, 70 percent, or 7 countries, were not free (Russia, Iran, Iraq, Saudi Arabia, Angola, Nigeria, and Sudan) and that 10 percent, or 1 country (Venezuela), was partly free in 2007. Further, only 20 percent, or 2 countries (Canada and Mexico), were free in 2007. Canada, Saudi Arabia, and Mexico are respectively the first, second, and third largest producers of oil for the United States, while China imports its largest percentages of oil from Saudi Arabia, Angola, and Iran, with less than 3 percent from free or partly free countries.

Moreover, in terms of energy resource competition between China and the United States, it appears that both China and the United States imported more than 50 percent of their oil from the same major producers in 2007 (Table 1.3). Those oil-producing countries were Russia, Iraq, Saudi Arabia, Angola, Nigeria, Canada and Venezuela. Of these, only Canada does not exercise state (or public) control over oil resources.

In light of the fact that both China and the United States import the majority of their oil and petroleum from socially and politically unstable countries—which potentially could result in supply disruptions, national economic disaster, and intensive competition for energy resources—it is necessary to determine empirically how and to what degree oil will influence this process, using a reliable forecasting methodology.

Table 1.3 Freedom in the World Compared with Annual Oil Production and Oil Imports by the United States and China in 2007

Country of Origin	*Freedom House Score*	*Annual Oil Production*	*Oil Exports to the People's Republic of China (bbl)*	*% of Total Annual Imports to China*	*Oil Exports to the United States (bbl)*	*% of Total Annual Imports to the US*
Russia	Not Free	3,604,022	106,549	8.91	151,074	3.07
Iran	Not Free	1,472,354	150,619	12.60	0	0
Iraq	Not Free	765,222.50	10,337	0.86	176,709	3.59
Saudi Arabia	Not Free	3,740,402	192,886	16.12	541,987	11.02
Angola	Not Free	645,567.70	183,115	15.30	185,352	3.77
Nigeria	Not Free	858,569	6,565	0.54	413,932	8.4
Sudan	Not Free	170,155.70	75,584	6.31	0	0
Venezuela	Partly Free	974,625	30,231	2.52	496,684	10.1
Canada	Free	1,249,026	3,442	0.29	895,976	18.2
Mexico	Free	1,277,535	0	0	559,304	11.37
Total (2007)		14,757,479	759,328	63	3,421,018	70

Source: Freedom House, Energy Intel (China) and U.S. Department of Energy (U.S.)

INDIA'S STRATEGIC ROLE

Although the potential impact of India's increased demand for imported crude oil and petroleum is not explicitly addressed in the three energy scenarios presented in this book, it is necessary to discuss whether India's demand for foreign oil resources will influence China's insatiable demand for foreign oil resources. In other words, will India's foreign oil demand intensify China's foreign oil demand given that India's economy has been growing at over 8 percent per year? At first look, a comparison of India's oil production versus oil consumption (in thousands of barrels per day) from 1980 to 2004 (see Figure 1.4) appears to indicate that the rate of growth in India's oil consumption has increased sharply relative to its oil production. However, when China and India are compared, it appears that China's rate of growth in oil consumption and production (in thousands of barrels per day) is much sharper than India's

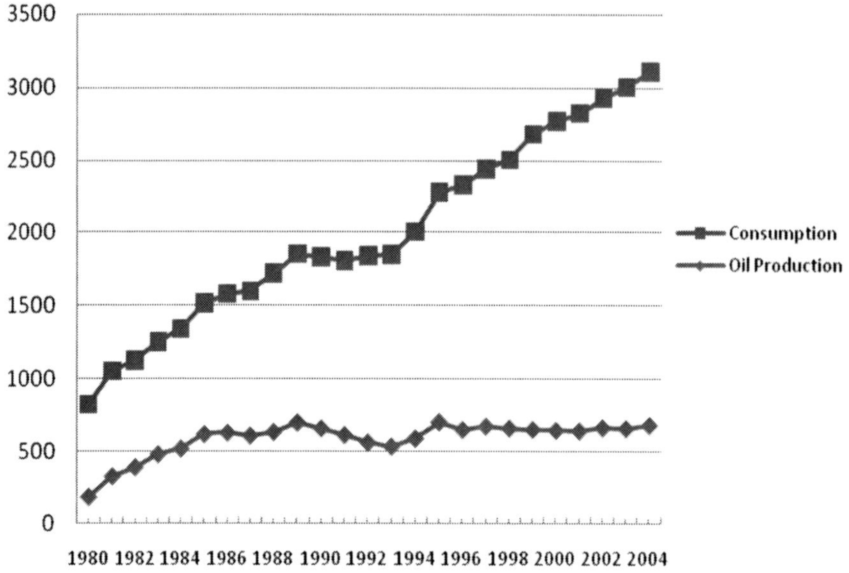

Source: Energy Information Agency, U.S. Department of Energy

Figure 1.4 India's Daily Oil Consumption vs. Production (1980–2004)

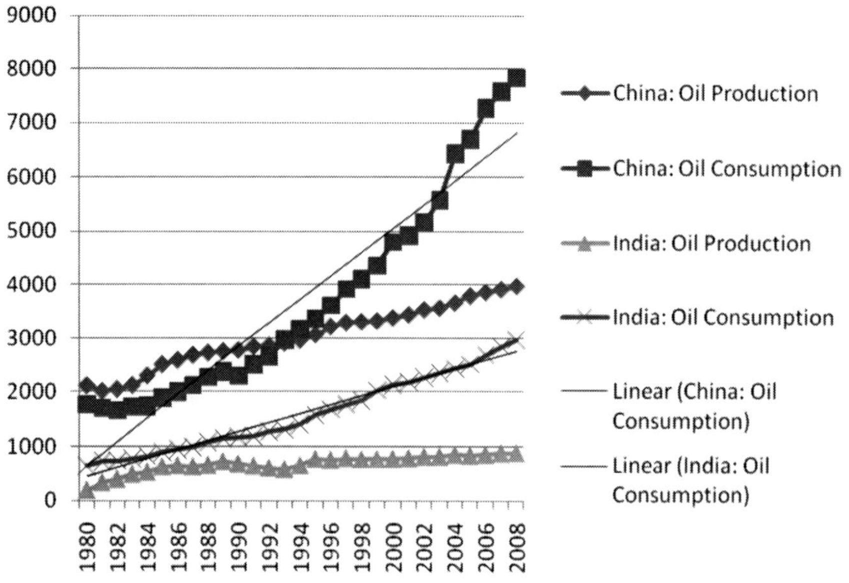

Source: Energy Information Agency, U.S. Department of Energy

Figure 1.5 A Comparison of China and India's Daily Oil Consumption vs. Production (1980–2008)

rate of growth in oil consumption and production (see Figure 1.5). Thus, India's foreign oil demand will not intensify China's foreign oil demand based on the fact that China's rate of growth in oil consumption and production historically has been considerably sharper than India's rate of growth in oil consumption and production.

Notes

1. Nanto, Dick K., James K. Jackson, Wayne M. Morrison, and Lawrence Kumins, *China and the CNOOC Bid for Unocal: Issues for Congress* (Washington, D.C.: Congressional Research Service, the Library of Congress), RL33093, February 27, 2006, 2.

2. Courtney, Hugh, Jane Kirkland, and Patrick Viguerie, "Strategy under Uncertainty," *Harvard Business Review*, November-December 1997, 68–71.

3. British Petroleum, *Energy Trends and Technologies for the Coming Decades: Address to the Harvard University Center for the Environment*, BP, March 8, 2006.

4. British Petroleum, *Energy Trends and Technologies for the Coming Decades: Address to the Harvard University Center for the Environment*, BP, March 8, 2006, 12.

5. Klare, Michael T., "Petroleum Anxiety and the Militarization of Energy Security," in Daniel Moran and James A. Russell, eds., *Energy Security and Global Politics: the Militarization of Resource Management* (New York: Routledge, 2009), 44.

6. Klare, Michael T.,"Petroleum Anxiety and the Militarization of Energy Security," 46.

7. Root, Hilton L., *Alliance Curse: How America Lost the Third World* (Washington, D.C.: The Brookings Institution), May 1, 2008, 4.

Part I

The Dependency Scenario

Chapter 2

The Competitive Dependency Scenario

BACKGROUND

The competitive dependency scenario is a future in which the United States and China are competing for crude oil and petroleum products internationally from the same oil-producing countries, and China's oil consumption exceeds oil production domestically. Conventional oil production has peaked in the United States, while conventional oil production in China is nearing its peak. This represents the situation as it exists in the first decade of the twenty-first century.

Since 1993 China has imported crude oil in order to sustain economic growth and development. Moreover, in recent years, China has imported over 1,196,270,000 barrels of crude oil annually of which 80 percent or more is imported from the same oil-producing states as the United States. Further, 90 percent of China's oil is imported by tankers originating from the Arabian Gulf and Sub-Saharan Africa. In addition, approximately 90 percent of China's oil tankers transit from the Indian Ocean to the South China Sea through the Straits of Malacca, a key strategic chokepoint in Southeast Asia, which potentially subjects China to supply disruptions caused by piracy, international terrorism and interdiction by U.S. naval forces.[1]

To avoid supply disruptions, China has developed diversification strategies like building oil pipelines from Russia, Central Asia, South Asia and Southeast Asia to China and has made port facility agreements with Burma and

Pakistan. Moreover, China has modernized its navy and facilitated bilateral dialogues with its neighbors in South Asia and Southeast Asia regarding the deployment of Chinese warships from naval bases in Burma, Bangladesh, Sri Lanka, the Maldives and Pakistan to protect the vital sea lines of communication (SLOCs) from the Straits of Hormuz and Horn of Africa to the South China Sea. Minimizing supply disruptions by ensuring the safe passage of China's oil tanker fleet transiting through the Indian Ocean region is critical to China's energy security because more than 50 percent of China's imported oil originates from the Near East and Sub-Saharan Africa.[2] Otherwise, China will not able to sustain economic growth and development as well as maintain Communist Party rule.

This scenario consists of multiple independent, national and events or issues of Chinese crude oil and petroleum exploration and acquisition that are currently occurring, have occurred, and may occur based on the most current literature available and oil and petroleum energy statistics collected for the years 2005, 2006 and 2007. Note that an analysis of the most recent statistics provides specific trends as to which states will matter in China's energy security in the future.

PROBABILITY OF OCCURRENCE

The overall probability of occurrence in the competitive dependency scenario is 92 percent. Regionally, the probability of occurrence is as follows: Asia-Pacific, 93 percent; Central Eurasia, 94 percent; the Middle East and North Africa (the Near East), 94 percent; Sub-Saharan Africa, 93 percent; Western Europe, 93 percent; and the Western Hemisphere, 94 percent. Therefore, regionally or overall, the competitive dependency scenario is a very probable scenario.

MACRO-STRATEGIC IMPLICATIONS FOR THE UNITED STATES

In this scenario, "China makes efforts to control its supplies from the oil wells all the way to China's coast, signing exclusive agreements with producer countries, buying or building oil exploring and trading companies, and developing strategies to defend the oil routes or to develop land-based lines of supply."[3] China's dependence on and pursuit of global energy resources causes it to be a strategic energy resource competitor, politically, economically, and militarily, to the United States in those countries where both nations obtain their petroleum products. The United States may be less politically influential globally over its direct oil supplier or indirect oil conduit nations, while China's global political influence may grow, especially in those countries that have control over the majority of the world's proven oil reserves such as the Organization of Petroleum Exporting Countries (OPEC). OPEC controls approximately 76 percent of the world's oil supply according to the *BP Statistical Review of World Energy* dated June 30, 2010. In this respect, China's pragmatic, nonideological approach to bilateral relations provides foreign governments with an alternative partner in international affairs, thereby reducing American influence in many countries, including the oil-producing nations.[4] U.S. allies in the Middle East are looking to diversify their diplomatic relationships away from their heavy dependence on the United States.[5]

Political competition derives from delinking political requirements from both oil trade arrangements and development assistance provided through loans, grants, and direct investments. In this scenario, China is able to compete successfully with the United States for global energy resources, complicating or obstructing U.S. political policies and actions taken against the oil-producing states, including requirements for political reforms, human rights improvements, and labor issues such as sweat shops and child labor.[6] Due to these policy linkages, U.S. dependence on foreign oil weakens its international leverage and entangles America with unstable and hostile regimes.[7] China, on the other hand, does not link policy issues to commercial oil ventures, making it more attractive as a partner and strengthening its global influence. China's pragmatic, nonideological approach to bilateral relations provides foreign governments—including the oil supplier governments—with an alternative to dependence on the United States, thereby reducing U.S. influence in many countries, including the oil supplier countries.[8] The high probability (92.2 percent) is that China will successfully out-compete the United States in some markets.

Economic competition derives from trade competition between the United States and China in the form of energy trading blocs between the oil-producing states and Chinese oil companies. If energy trading blocs favor China's national oil companies over the independent oil companies (IOCs) operating in politically nonaligned countries in the developing world, America's ability to acquire cheap oil may be undermined, especially if China is able to use lucrative infrastructure investments to entice oil suppliers to favor China's national oil companies over independent oil companies in exploration and production contracts. This action could effectively shut out U.S.-owned oil companies like Exxon Mobil and Chevron. The United States depends on corruption-free access to reserves so that American companies can compete on a level playing field.

China's pursuit of global energy resources influences the price of crude oil and petroleum in the United States. The United States Congressional Budget Office (CBO) stated in its 2006 report, *China's Growing Demand for Oil and Its Impact on U.S. Petroleum Markets*:

> World oil prices have been rising since 2004, driven in part by the increases in China's demand for crude oil and refined petroleum products. Those increases were most prominent for the light petroleum products—gasoline and diesel—that are used primarily for transportation. Over the next five years, the pace of growth in China's demand for oil in general, and for transportation fuels in particular, could be a key factor contributing to further increases in the prices for crude oil and refined petroleum products. In the United States, those price increases could affect the demand for petroleum and influence the investment decisions of U.S. refiners.[9]

The United States also faces potential military competition from China over energy resources in the near future. China's defense industry and its military (PLA, or the People's Liberation Army) find opportunities under this scenario to advance its foreign policy interests and improve its military relations with the oil-producing nations because of Beijing's pragmatic, nonideological approach to its foreign relations. As the United States has often done, China can further its military relations with the oil-producing nations in the wake of successful diplomatic and commercial ties which, in turn, may have serious military consequences for the United States, such as resorting to various military measures to protect its energy assets, notwithstanding its accession to international conventions such as the Law of the Sea. These measures include direct seizure of energy assets by military means; destruction of energy assets to deny their use to rivals like the United States or India; military competition arising from competitive efforts to exploit energy resources either on land or sea; military protection of, or attacks upon energy production and transportation infrastructure, including oil fields, refineries, pipelines, and port facilities; conveyance of major military assets to regional energy producers in exchange for a preferential market; and active military control of military straits through which energy assets move.[10]

It is important to note that Chinese foreign military operations constitute a strategic-level activity in support of the larger, foreign, diplomatic, economic, and security agenda established by China's leadership.[11] David H. Shinn opined that:

> The PLA does not engage in freestanding military initiatives conducted by military professionals for military reasons. In other words, the PLA is not an

independent actor; it must coordinate its activities with the party and the state bureaucracy. The PLA maintains an ambitious global program of military exchanges and training programs.[12]

Overall, China's demand for oil may shape or influence its military policies and may complicate any United States efforts for preserving international stability and cooperation through military means.

As well as the macro strategic implications for the United States, it is essential to explore and discuss how specific geographic regions will comparatively affect the United States politically, economically, and militarily, because the United States and China have varied levels of political, diplomatic, economic, and military relations with and influence over energy suppliers. Further, any scholarly work on global energy security, especially where the United States may be affected by Chinese political, economic or military relations or influence, would not be credible without appreciating the regional differences. Therefore, regional analyses are made in this energy scenario for a greater understanding of how each region specifically affects U.S. oil energy security.

REGIONAL ANALYSES

In the competitive dependency scenario, regional analyses of China's global oil trade and its implications for the United States are important in order to determine how United States oil energy security is affected. They are the central focus of PRINCE analysis because as a risk analysis assessment method, the PRINCE model quantitatively determines whether China and the United States will peacefully share the world's oil reserves or whether competition for crude oil and petroleum reserves will lead to conflict. (See Appendix B for a description and mechanics of the PRINCE model.)

Realistically, the impact of China's global oil trade on the United States will vary (positively, negatively or none) based on the extent of overall U.S. and Chinese political, economic/trade and military relations, aid and influence and amount of oil energy resources available in each region. (See Appendix A for available oil resources.)

The available hydrocarbon resources, proven crude oil reserves as of January 1, 2009 are shown on page 20 according to the *Oil and Gas Journal.* (See Table 2.1).

The Near East and the Western Hemisphere hold the majority of proven oil reserves, 60 percent and 25 percent respectively, and are clearly the most important regions of the world in terms of energy security for both the United States and China. This book focuses not only the Near East and the Western

Table 2.1 Proven Oil Reserves

Region	*Billion Barrels*	*% of World Total*
Russia	60	4.47%
Asia-Pacific	34	2.53%
Central Eurasia	38.88	2.90%
Near East (Middle East & North Africa)	805.98	60.00%
Africa (Sub-Saharan Africa)	57.07	4.25%
Western Hemisphere (North & South America)	332.59	24.78%
Europe	13.66	1.02%
Total	1,342.07	100%

Source: Oil and Gas Journal

Hemisphere but also Russia, the Asia-Pacific, Central Eurasia, Sub-Saharan Africa and Western Europe, where Chinese and U.S. oil interests overlap. Key oil-producing states such as Angola, Canada, Iran, Nigeria, Saudi Arabia, Sudan, and Venezuela are examined.

Chapter 3 examines China, Russia, Asia-Pacific and South Asia and explores China's geopolitical ambitions in the South China Sea and the Indian Ocean.

Chapter 4 looks at Central Eurasia, including the oil pipelines into China.

Chapter 5 focuses on the Middle East and North Africa, including China's dangerous relationship with Iran.

In Chapter 6, Sub-Saharan Africa is examined, including China's emerging strategic partnerships in the region, including Angola, Nigeria, and Sudan.

Chapter 7 describes China's existing, albeit inconsequential, oil energy relationships with Norway and the United Kingdom in Western Europe.

Chapter 8 analyzes the Western Hemisphere, emphasizing China's burgeoning economic relationships with Canada and Venezuela.

Chapter 9 concludes the discussion and summarizes the key findings. It is no surprise that China's growing strategic engagement and rising geopolitical influence is problematic for the United States, especially in the Near East and Sub-Saharan Africa.

Notes

1. Leverett, Flynt, "Resource Merchantism and the Militarization of Resource Management: Rising Asia and the Future of American Primacy," in Daniel Moran

and James A. Russell, eds., *Energy Security and Global Politics: the Militarization of Resource Management* (New York: Routledge), 2009, 217.

2. Alterman, Jon. B. and John W. Garver. *The Vital Triangle: China, The United States and the Middle East* (Washington, D.C.: CSIS Press), 2008, 6.

3. Stanzel, Volker, "The EU and China in the global system," in David Shambaugh, Eberhard Sandschneider and Zhou Hong, eds., *China-Europe Relations: Perceptions, Policies and Prospects* (New York: Routledge), 2008, 258.

4. Saunders, Philip C., *China's Global Activism: Strategy, Drivers, and Tools* (Washington, D.C.: Institute for National Strategic Studies, National Defense University), 2006, 28.

5. Lieberthal, Kenneth and Mikkal Herberg, *China's Search for Energy Security: Implications for U.S. Policy* (Seattle: National Bureau of Asian Research), Volume 17, Number 1, April 2006, 24.

6. Saunders, *China's Global Activism: Strategy, Drivers, and Tools,* 28 and Bo Kong, *An Anatomy of China's Insecurity and Its Strategies* (Seattle: Pacific Northwest Center for Global Security), PNNL-15529, December 2005, 55.

7. Military Advisory Board, *Powering America's Defense: Energy and the Risks to National Security* (Alexandria, VA: Center for Naval Analysis), May 2009, 1.

8. Saunders, *China's Global Activism: Strategy, Drivers, and Tools,* 28.

9. U.S. Congress, Congressional Budget Office, *China's Growing Demand for Oil and Its Impact on U.S. Petroleum Markets* (CBO), April 2006, 1.

10. Moran, Daniel and James A. Russell, "Introduction: the militarization of energy security," in Daniel Moran and James A. Russell, eds., *Energy Security and Global Politics: the Militarization of Resource Management* (New York: Routledge), 2009, 7.

11. Shinn, David H., "Military and Security Relations: China, Africa, and the Rest of the World" in Robert I. Rotberg, ed., *China into Africa: Trade, Aid, and Influence* (Washington, D.C.: Brookings Institution Press), 2008, 162.

12. Shinn, David H., "Military and Security Relations: China, Africa, and the Rest of the World," 162.

Chapter 3

The Asia-Pacific

The Asia-Pacific region is made up of North Asia (Russia), East Asia, Southeast Asia including Oceania (Australia), and South Asia. Many observers see this region as a crucial area of Sino-American competition because the United States has deep security, trade, and investment interests with the region.[1] United States oil energy trading, crude oil and petroleum imports from the Asia-Pacific are relatively low compared to the other resource-rich regions of the world.

Similarly, the Asia-Pacific is becoming less significant for China in terms of hydrocarbon energy trading. China has imported less crude oil and petroleum from the Asia-Pacific than any other region except Europe during the last since 2005. Many Asia-Pacific oil exporting nations, especially those in Southeast Asia, are gradually approaching a decline in their hydrocarbon energy reserves. Thus, the Asia-Pacific is not really a "battleground" region for Sino-American hydrocarbon competition.

Although China imports less crude oil and petroleum from the Asia-Pacific than any other region in the world, the Asia-Pacific is not any less important in terms of energy security. Approximately 93.5 percent of China's crude oil and petroleum imports are transported by sea using ocean-going tankers and transit through either the Straits of Hormuz or Straits of Malacca, two major strategic chokepoints that are constantly patrolled by the United States Navy.

Because the PLA Navy has neither the military nor logistical capability to patrol these vital areas and challenge U.S. naval supremacy, China has to rely on means other than military power to ensure the steady and uninterrupted flow of oil from oil rich regions of the world.

One way for Beijing to ensure the safe transit of its foreign oil was to become more regionally involved through trade, political dialogue, and cultural exchanges. While China appears to be genuinely benevolent, most Asia-Pacific nations are wary of its true intentions for the Asia-Pacific region. Yet, many Asian-Pacific nations have engaged China on various levels, which has enhanced their relationship with the Middle Kingdom. For the most part, China's soft power approach (through cultural and economic means), or so-called "charm offensive," has paid enormous dividends in recent years, resulting in stronger political, economic and cultural ties. However, such ties bear closer scrutiny because of variations in actual political, economic, and military relations between China and the other nations of the Asia-Pacific region.

NORTH ASIA (RUSSIA)

The PRINCE model predicts a 94.1 percent probability that an oil pipeline will be built from Siberia into Northern China and a 97.9 percent probability that China and Russia will jointly explore for oil in the Arctic Circle in the near future.[2]

Political, Economic, and Military Implications for the United States

China's energy needs, along with those of the rest of the large Asian oil and gas importers, are also inexorably drawing in Russia back into Asia as a key strategic and commercial player with a range of potential implications for both U.S. interests in Asia and U.S. relations with Russia. Energy has the potential to strengthen long-term political ties between China and Russia, but this has not occurred largely due to Russian fears over China's growing regional power and fears that Russia's Far East region may be overrun in the future by Chinese power.[3] Consequently, in spite of some strategic energy agreements that made between the two countries in the last four years, Russia has frustrated China's efforts to forge new major energy deals.

The most glaring example of a row in Sino-Russian energy relations is Russia's 2005 decision to build the East Siberian oil pipeline to the Pacific Coast rather than live up to its previous agreement with China to build the oil pipeline to Northeastern China. The first leg of the Siberian oil pipeline would run 1,713 miles from Taishet to Skovorodino (in the Amur Region of Russia's Far East) with an expected completion date of December 2008, and the second leg would continue for 1,304 miles from Skovorodino to the Perevoznaya Bay on Russia's west coast. Originally, the second leg was supposed to run from Skovorodino to the Chinese city of Daqing. However, Russia delayed the project until it signed an agreement with China on October 28, 2008. Arguments between China and Russia as to which country should pay for this Sino-Russian leg, Russia's economic hedging of Japan against China, Russia's need for security and diversification of the energy market and Russia's potential economic opportunities in the Asian oil market have forced Russia to reconsider the project. In direct competition, Japan was offering twice as much as the Chinese to build the pipeline—US $14 billion—as well as offering to fund the construction of a refinery and oil-loading port on the eastern Siberian coast. [4]

In terms of future strategic energy cooperation, China's largest oil company—China National Petroleum Corporation (CNPC)—and Rosneft, one of three major state-run Russian companies, may jointly explore for oil in the Kara Sea in the Arctic Circle in the near future. Oil exploration in the Kara Sea would be a major energy coup for both nations but even more so for Russia, because it could elevate its position from the world's second largest oil exporter to the first. Thus, while potentially promising, this event is highly speculative because no agreements have been signed to date. However, with an event occurrence probability of 97.6 percent, there is a very good chance that this event may occur in the near future. This would give China greater oil energy security, given the geographic location of the event and its geographic inaccessibility to other major oil regions of the world.

Sino-Russian oil energy relations are unlikely to have any serious political consequences for the United States because of past Sino-Russian tensions over energy cooperation. China is more likely to be affected in terms of energy security because it is more oil energy dependent on Russia than the United States is. In this respect, 11 percent of China's total crude oil and petroleum imports are from Russia, compared to the United States' total crude oil and petroleum imports of 3 percent. Despite China's energy insecurity, it is unlikely that the Russo-American relations will improve because of political issues unrelated to oil energy, such as NATO expansion into Eastern Europe, the U.S. presence in the Middle East, and perceived American hegemony in Central Eurasia in the wake of the September 11 attacks. Further, in light of recent Russian military presence in and major military equipment sales to Venezuela, which is in America's backyard, it is more likely that Russo-American relations will be tepid.

Similar to China's political relations, a strong Sino-Russian oil energy partnership will not necessarily adversely affect the United States economically either regionally or globally because the United States does not depend on Russia for the majority of its foreign trade, including crude oil and petroleum.

In this respect, Russia exports annually approximately 4 percent of America's total crude oil and petroleum and is not one of America's top fifteen trade importers or exporters. Canada and China are America's top import-export trading partners. Further, Russia seeks to diversify its hydrocarbon energy market by making its crude oil and petroleum available not only to China but also to Japan and South Korea. Russia recently decided to redirect the construction of the East Siberian oil pipeline to Russia's Pacific Coast rather than to Daqing in Northern China. In addition, tenuous Sino-Russian energy relations have not made for improved Sino-Russian political relations, even with Dmitri Medvedev in power. Therefore, it is highly unlikely that the United States will be impacted economically.

Similar to political and economic relations, a strong Sino-Russian oil energy partnership will not necessarily affect the United States militarily either regionally or globally or lead to a military competition for strategic energy resources because the United States, unlike China, does not depend on Russia for military weapons, munitions, or equipment. Although China and Russia have neither a strong energy partnership nor an existing military alliance, if a strong Sino-Russian oil energy partnership facilitates the development of a Sino-Russian military alliance, then serious military consequences could ensue for the United States, especially in the Asia-Pacific region. Such a military alliance could upset the balance of power in the Asia-Pacific region, potentially derailing any further Six-Party talks regarding the North Korean nuclear program and resulting in a greater Sino-Russian voting coalition in the United Nations Security Council. It could also put the U.S. military on notice in East Asia with regard to the sea lines of communications (SLOCs) that connect these vital international trade zones.

However, China and Russia do not have a military alliance. Rather, China and Russia have enhanced security cooperation through cultural exchanges, joint military exercise participation, and membership in the Shanghai Cooperation Organization (SCO), which is focused on combating extremism, separatism, and terrorism in Central Eurasia. It should be noted that China has never been a member of a formal military alliance.

EAST ASIA

The PRINCE model predicts that China will continue to explore for oil domestically by 91.4 percent, in the littorals by 97.5 percent, and in the East China Sea by 71.4 percent and will continue to import oil from Mongolia and South Korea by 97.65 percent and 98.1 percent respectively.[5]

Political, Economic, and Military Implications for the United States

China's hydrocarbon energy trading with East Asia (Japan, Mongolia, and South Korea) is unlikely to shift the balance of political power from the United States to China because of the strong diplomatic relationships that exist between the United States, Japan, and South Korea. Hydrocarbon energy trading does not currently shape or influence diplomatic relations in East Asia. Normal Sino-Mongolian diplomatic relations range from cordial to warm, while diplomatic relations with Japan and South Korea are respectively tepid and warm.[6] Conversely, U.S. diplomatic relations with Mongolia, Japan, and South Korea are respectively warm, hot, and very warm.

Similar to political relations, China's hydrocarbon energy trading with East Asia (Japan, Mongolia, and South Korea) does not play a pivotal role in shaping or influencing economic relations and is unlikely to shift the balance of economic power from the United States to China because of the strong economic relationships (U.S.-China-Mongolia, U.S.-China-Japan, U.S.-China-South Korea) that exist between the five nations. International trade matters in East Asia provide for greater regional stability than diplomatic and military relations: Sino-Mongolian economic relations are warm, while China's economic relations with both Japan and South Korea are hot.[7] Similarly, U.S. economic relations with Mongolia, Japan, and South Korea are respectively warm, hot and hot.

Similar to the political and economic relations, China's hydrocarbon energy trading with East Asia (Japan, Mongolia. and South Korea) is unlikely to shift the balance of military power from the United States to China or lead to a military competition for energy resources because of the close military relationships that exist between the United States, Japan, and South Korea. In this respect, hydrocarbon energy trading does not play a role in shaping or influencing military relations between the United States, China, Mongolia, Japan and South Korea. In terms of military relations in East Asia, Sino-Mongolian military relations are cordial, while China's military relations with Japan and South Korea are respectively cold and cool.[8] Conversely, U.S. military relations with Mongolia, Japan, and South Korea are respectively warm, hot, and hot.

The Political, Economic, and Military Implications of China's Domestic Hydrocarbon Opeations

Domestic hydrocarbon operations (exploration, discovery, extraction and refinement) in territorial China—such as the Tarim Basin in Xinjiang, Bohai Bay and coastal China—are unlikely to have any negative diplomatic, economic or military consequences for the United States such as political, economic, or military competition for strategic energy resources. These operations are sovereign in nature and do not involve the United States or any other country, except in a capacity where technologically advanced countries like the United States may be able to provide public and private technical expertise to assist China in acquiring its natural hydrocarbon energy resources from underground and the below the ocean floor. If China conducts domestic hydrocarbon operations and seeks greater international cooperation from the United States through the public and private technical assistance, then both countries will benefit on many levels, such as increasing their strategic petroleum reserves (SPR) as well as lowering global oil prices. Therefore, it behooves both China and the United States to engage each other and work constructively to help reduce China's dependency on foreign hydrocarbons.

SOUTHEAST ASIA INCLUDING OCEANIA

The PRINCE model predicts that China will continue to explore for oil in and import oil from Southeast Asia based on a probability of 85 percent or greater. However, it is uncertain as to whether China will unilaterally explore, discover, extract, and refine crude oil from the Spratly Islands based on a probability of 58.1 percent.[9]

Political, Economic, and Military Implications for the United States

China's hydrocarbon energy trading with Southeast Asia [Australia, Brunei, Burma (Myanmar), Indonesia, Malaysia, Philippines, Thailand and Vietnam] is unlikely to shift the balance of political power from the United States to China because of the strong diplomatic relationships that exist between the United States and the Southeast Asian countries with the exception of Burma. "Both China and the United States have strong ties to Southeast Asia, and both draw upon considerable strengths in projecting soft power in the region."[10] Chinese soft power is implemented in the form of aid, investment, and friendship, but many Southeast Asian countries are not anxious for China to dominate the region militarily. Southeast Asian nations deeply value the U.S. security umbrella against a potentially expansive China despite widespread improvements in public perceptions of China and parallel declines in perceptions of the United States.[11] The United States "has the added advantage of being seen by Southeast Asian and other Asian government leaders as Asia's 'least distrusted power' as these leaders seek to enhance their nationalistic ambitions and independence of action among other Asian governments, including and especially China, that have territorial and other ambitions directly at odds with their national interests."[12] The strategic importance of Southeast Asia as the site of future oil supplies is low. "Southeast Asia simply does not have the potential for future oil supplies to be a major part of the solution to China's security angst,"[13] causing it to remain a sideshow at least insofar as oil supplies are concerned. However, Southeast Asia will maintain its strategic importance to China because of the regional sea lines of communication (SLOCs). "China's growing influence derives mainly from its role as a market for the region's natural resources," even though the region contains approximately 3.3 percent of the world's proven oil reserves, while the United States "maintains its influence based on its military presence, foreign direct investment, its market for the region's manufactured goods, military and development assistance and educational opportunities."[14]

However, China's influence and hydrocarbon energy security in Southeast Asia could be set back by disruptions in the sea lines of communication such as a sovereignty dispute in the South China Sea that threatens military conflict; terrorism and international crime, usually falling under the rubric of "piracy" in the Malacca Strait[15]; and a U.S. Naval blockade in the Malacca Strait in the event that China militarily invades Taiwan. Since 60 percent of China's crude oil imports currently pass through the Malacca Strait and the South China Sea with an estimated increase towards 75 percent by the year 2015,[16] the safety and security of these sea lines of communication becomes a critical factor not only to China's regional influence and hydrocarbon energy security but also its economic growth and continued rule by the Chinese Communist Party, because "the threat of economic stagnation (caused by maritime trade disruptions) raises real risks of social instability, which could in turn threaten the continued political monopoly of the Chinese Communist Party."[17] Since oil imports destined for China from Africa and the Arabian Gulf are vulnerable to terrorism, piracy, U.S. naval interdiction in and around the Indian Ocean and the Strait of Malacca, and potential military conflict in the South China Sea, it behooves China to pursue various hydrocarbon energy security strategies that will reduce its perceived vulnerabilities associated with its maritime oil imports.

China is considering, among various hydrocarbon energy security strategies, one that is based on the diversification of energy import routes through the development of Pakistani and Burmese transnational pipelines. Beijing is also considering the idea of building a shipping canal through Thailand's Kra Isthmus to divert her maritime fleet directly into the Gulf of Thailand. As a direct consequence of the growing volume of crude oil flowing to China by tanker through the Indian Ocean and South China Sea, China is developing its naval capability to reach beyond coastal waters to the open ocean or the blue water, whereby naval assets will be used to its protect global maritime shipping interests (including the oil tankers) transiting through Indian Ocean and elsewhere and will include large capital ships like a nuclear aircraft carrier (CVN) and Russian-made diesel submarines.[18] This notion is known as the militarization of resource management resulting from energy insecurity.

Further, in tandem with the resource militarization Herberg opined, "China is developing naval port access agreements along the coasts of the Indian Ocean, most importantly in Pakistan, Myanmar (Burma) and Bangladesh, that would allow its ships and submarines to operate more effectively there and into the Strait of Malacca."[19]

This strategy is appropriately named the "string of pearls strategy" by a Pentagon report. "The 'String of Pearls' describes the manifestation of China's rising geopolitical influence through efforts to increase access to ports and airfields, develop special diplomatic relationships, and modernize military forces that extend from the South China Sea through the Strait of Malacca, across the Indian Ocean, and on to the Arabian Gulf."[20]

> In addition, Herberg opined, "China is rapidly accelerating its own submarine development along with the acquisition of sophisticated Russian-made submarines capable of operating long distances."[21]

China is already paving the way for this new submarine force by expanding its submarine-basing facilities on Hainan Island in the South China Sea.

Finally, China has embarked on a program to build its own tanker fleet to move its oil supplies because only 10 percent of its oil imports arrive by Chinese tankers.[22] It should be noted that strong diplomatic relationships are a must with the maritime nations of Indonesia, Malaysia, Philippines, Singapore, and Thailand because it is critical to China's energy security strategy in Southeast Asia, especially since these countries are geographically situated along major maritime trade routes.

The "'militarization' of resource management describes a growing inclination of energy producers and consumers to base decisions about their behavior in the global energy markets on political and strategic calculations as well as commercial and economic considerations."[23] In China's case, resource militarization evolves from a mercantilist energy policy that reflects its sense of weakness and vulnerability regarding reliable access to energy supplies.[24] This sense of weakness provides the rationale for direct state intervention and support from Beijing rather than plenipotentiary oversight by the three Chinese national oil companies (CNPC, CNOOC, and Sinopec). Further, Chinese energy policy, which provides the basis for resource militarization, is shaped by three factors that are mercantilist in nature: that there is a general mistrust of global energy markets, that distrust of the energy markets is aggravated by the perception that the energy markets are controlled by the United States, and that in terms of energy sector capabilities, Beijing feels that China is working from a position of weakness and must play catch-up. [25] Thus, China is an outstanding example of both mercantilism and resource militarization because its "rapidly growing navy is universally viewed as derivative of that nation's export-driven economic expansion, and corresponding dependence on raw materials imported by sea."[26] China's naval capability is becoming more Mahanian (global and strategic) in nature, growing in tandem with its sea-borne dependent economy and providing protection to its merchant fleet on the high seas while supplementing its continental strategy of energy import route diversification through Burma and Pakistan.

Similarly to political relations, China's hydrocarbon energy trading with Southeast Asia (Australia, Brunei, Burma (Myanmar), Indonesia, Malaysia, Philippines, Thailand, and Vietnam) is unlikely to shift the balance of economic power from the United States to China or make the United States less economically influential in the region. This is because of the strong commercial relationships that exist between the United States and the Southeast Asian countries, with the exception of Burma, even though China is a rising power seeking to both create a sphere of influence and to erode U.S. dominance in Southeast Asia. The United States continues to exert both hard and soft power throughout Southeast Asia.[27]

In terms of American soft power, the Congressional Research Service opined:

> The United States maintains multi-faceted foreign aid programs with clear objectives and large development and humanitarian components. The United States was also a major contributor to countries hit by the 2004 Indian Ocean tsunami, which affected several Southeast Asian countries. Further, the United States remains ASEAN's second largest trading partner (China ranks fifth) and its fourth largest source of foreign direct investment (China ranks tenth), and has sought free trade agreements with several countries in the region.[28]

The United States maintains a strong security presence (consisting of air, land, and naval assets) throughout Asia including Southeast Asia, where the United States and many ASEAN countries have bilateral security arrangements. Hence, the "China-Southeast Asia dyad is mediated by a strong Southeast Asian desire to sustain U.S. economic, political, and security interests as a means of externally balancing China's current dynamism."[29]

Beyond the goal of establishing a sphere of economic influence in Southeast Asia that is potentially inimical to the United States, China seeks to ensure its energy security to sustain economic growth and the longevity of the Chinese Communist Party. Mikkal E. Herberg opined that:

> As China's economy grows and its energy needs boom, energy supplies and transit are likely to become increasingly important factors in China's relations with Southeast Asia and the region's key players. In other words, the geopolitics of energy in the region are likely to become an important force in shaping overall regional relations.[30]

Herberg's comment raises questions as to how China's hydrocarbon energy trading with the countries of Southeast Asia have shaped economic relations vis-à-vis the United States in such a way that it has strategic implications of an economic nature. A more in-depth analysis of China's hydrocarbon energy trading with Southeast Asia is required to see what effect, if any, it has on the United States economically.

Similarly to political and economic relations, China's hydrocarbon energy trading with Southeast Asia (Australia, Brunei, Burma (Myanmar), Indonesia, Malaysia, Philippines, Thailand, and Vietnam) is unlikely to shift the balance of military power from the United States to China or lead to a military competition over strategic energy resources because of the strong military relationships that exist between the United States and the Southeast Asian countries with the exception of Burma, Indonesia and Vietnam. In this respect, the United States' strong military relations are a direct consequence of its strong political relations in Southeast Asia. It is the U.S. military presence, rather than that of China, that is seen as guaranteeing peace and stability in the region. China has not indicated a desire to assume this role.[31]

Chambers believes that China must develop greater economic capacity and be viewed as generating a greater positive political presence within the region, without its military appearing to be threatening, if it is to emerge as a great power within Asia.

China's military power in Southeast Asia is slowly evolving in the wake of its political/diplomatic power. Yet currently, its military relations are somewhat limited. Burma is China's only true military ally in the region, and it plays a significant role of China's hydrocarbon energy security strategy of route diversification and "string of pearls" strategy.

SOUTH ASIA

South Asia does not produce oil for either China or the United States. However, South Asia is critical to both Chinese and United States energy security because the region straddles the Indian Ocean and contains vital shipping lanes by which crude oil is transported by oil tankers to China and the United States from the Persian Gulf and Africa. It is of particular strategic importance to China because 30 percent (of China's oil is imported from Africa and approximately 50 percent is imported from the Persian Gulf region. Since China does not have the blue-water naval capability to protect its tanker fleet beyond the South China Sea, the responsibility lies with the U.S. Navy to patrol the sea lines of communication (SLOCs)—not only the Persian Gulf and the Indian Ocean but also the strategic chokepoints at the eastern and western approaches to the Indian Ocean at the Straits of Malacca, the Strait of Bab-el-Mandeb (between the Red Sea and the Gulf of Aden) and the Straits of Hormuz.

Military reliance on the United States Pacific Fleet to keep the sea lines of communication open from Africa, the Red Sea and Persian Gulf to the South China Sea via the Indian Ocean poses a strategic quandary for China because China's oil lifelines could be severed in the event of an East Asian crisis such as an attack on Taiwan. In this respect, China would be powerless to stop a U.S. naval blockade of any oil tankers bound for China that were transiting through the Straits of Hormuz or Straits of Malacca.

China will also have to contend with Indian naval power in the Indian Ocean region because India uses the same sea lines of communication, relying on these sea-lanes for 90 percent of its oil and over 90 percent of its foreign trade, competes globally with China for oil energy resources, and has the most capable navy in the region The Indian Navy, which consists of an aircraft carrier, fourteen submarines and fifteen major surface combatants, can project sea power not only in the Indian Ocean region but also in the South China Sea, the Mediterranean Sea, and the Persian Gulf. Its introduction of the Brahmos cruise missile and the possible lease of nuclear submarines from Russia will add significantly to the Indian Navy's flexibility and striking power in South Asia.[32] Given the Indian Navy's improved technology and military reach, Beijing views India as a strategic competitor, even though overall relations have improved considerably between the two nations since the end of the Cold War.

To counter Indian and United States naval influence in the Indian Ocean region China has quietly followed a strategy of military and economic cooperation with Pakistan, Myanmar, Bangladesh, Sri Lanka, and the Maldives.[33] China has been engaged in building a major naval base on Hangyi Island in Myanmar near the Irrawady river delta, which will be soon be capable of hosting large Chinese naval vessels like nuclear attack submarines (SSNs) and ballistic missile submarines (SSBNs).[34] Beijing has also been selling arms as well as conducting high-level talks about establishing a naval base with the government of Bangladesh, which would provide berthing facilities for the Chinese surface ships and submarines.[35] China is exploring the possibility of establishing an intelligence listening post in Sri Lanka as well as berthing facilities for its warships, and is engaged in negotiations with the Maldivian government to construct a submarine base at Marao Island.[36] Finally, in apparent contravention to its improved relations with the Indian government, China is building a deep-water port at Gwadar, Pakistan, that will not only provide China and Pakistan the ability to monitor the sea lanes from the Persian Gulf but also the ability to monitor U.S. naval activity in the Persian Gulf region and Indian naval activity in the Arabian Sea.[37] Berthing facilities can support a future Chinese Navy Carrier Battle Group (CVBG) as well as Chinese nuclear attack submarines (SSNs), ballistic missile submarines (SSBNs), and guided missile submarines (SSGNs). The port supplies a maritime interface to the Sino-Pakistani Friendship Karakorum Highway that opened China's access to Pakistan and southern Asia in the late 1960s.[38] Essentially, the port is a transshipment point for Chinese maritime trade. China can potentially transport crude oil to Western China via a future pipeline and thereby reduce its maritime vulnerability. Transit time and the costs of shipping crude oil from the Persian Gulf, notwithstanding the $5.9 billion cost to build the oil pipeline, will be reduced.[39] Thus Gwadar offers a geo-economic and geo-strategic pivot to China and Pakistan.[40]

However, the realization of economic and strategic objectives of the Gwadar port is largely dependent on upon the reduction of separatist violence in Baluchistan and Xinjiang. Baluch nationalists see the Gwadar port project as an attempt at colonization and the capture of their natural resources.[41]

Overall, China's developing military and economic cooperation with the nations of South Asia will make it better positioned to protect its maritime energy transport routes during any future crisis. Leiberthal and Herberg state that "this cooperation, in turn, risks exacerbating the broader tensions between the United States and China over the increasing pace and scale of China's military and naval modernization."[42] Beijing's developing relations and cooperation with the nations of South Asia will have strategic consequences for the United States.

The PRINCE model predicts that China will build and operate oil pipelines from Russia to India based on a probability of 89 percent and oil refineries in Pakistan based on a probability of 61 percent. Further, the model predicts that China will develop military ties with India based on a probability of 69 percent.[43]

Political, Economic and Military Implications for the United States

Washington seeks effective engagement and close regional cooperation, to defuse the Indo-Pakistan crisis, promote cooperation in the war on terrorism, and strengthen democratic institutions,[44] while China seeks to expand multidimensional cooperative relations with all countries in the region.[45] As explained above, China is building strategic relationships and developing a capability to establish a forward presence along the sea lines of communication in the Indian Ocean region that connect China to Africa and the Near East. Beijing continues to assert its sovereignty over Taiwan, countering Taiwan's efforts to expand its diplomatic ties regionally and with the United States, while working to guarantee South Asian neutrality in the event of a Sino-American military clash over Taiwan.[46]

It also seeks to minimize India's political alignment with the United States by encouraging U.S. engagement with Pakistan.[47] With Gwadar as a potential berthing location for Chinese warships, a Chinese naval base 240 miles from the Straits of Hormuz would provide not only strategic depth for China in the Indian Ocean region but may also directly threaten U.S. interests in the region. However, warming diplomatic ties between China and Pakistan will not necessarily occur at the expense of U.S.-Pakistani relations in the short term because the United States regards Pakistan as an ally in its war on the Taliban and al Qaeda.

Pakistan has long figured in U.S. strategic interests in the region. It was a member of the Baghdad Pact, CENTO, and SEATO during the Cold War, and facilitated talks between the United States and China in 1969, which resulted in the full normalization of diplomatic relations in 1972.[48] Currently the United States supplies Pakistan with financial and material support for the U.S.-led global war on terror in South Asia.

U.S. economic interests in South Asia are primarily to strengthen its economic and trade ties with South Asia as well as to promote economic growth for every state in South Asia.[49] China seeks to strengthen its economic and trade relations with South Asia, especially with India. Strengthening economic and trade relations with India may help to reduce the lingering mistrust between the two nations with respect to past border disputes and regional power rivalry.

Sino-Pakistani economic and trade ties, which are stronger than Sino-Indian economic and trade ties, are nurtured within the framework of military/strategic needs.[50] While China has been one of Pakistan's major suppliers of military aid and equipment, its energy activities in South Asia are unlikely to result in a Sino-American economic competition for oil energy resources because the region does not produce crude oil for either the United States or China. However, South Asia's economic significance is no less important to the United States and China because of its geographic location relative to the SLOCs. In this respect, South Asia is vital to both U.S. and Chinese energy security. The PRINCE analysis indicates that both the United States and China will seek to minimize their economic risk along the SLOCs through continued military and economic cooperation with the nations of South Asia.

To establish regional security and stability, the United States conducts military operations in support of political objectives such as counterterrorism and WMD non- and counterproliferation. It also supports regional development and cooperation by providing military equipment, training and assistance, and information sharing, as well as promoting military outreach and exchange programs with the objective of building stronger relationships with regional military players such as India and Pakistan.

To sustain international energy security, the U.S. Navy provides for the safe passage of national and international oil tankers transiting in the Indian Ocean region including the strategic chokepoints at the eastern and western approaches at the Straits of Malacca, the Strait of Bab-el-Mandeb (between the Red Sea and the Gulf of Aden), and the Straits of Hormuz. International navies from Asian and European countries are working in concert with the U.S. Navy to patrol at these strategic chokepoints. China has shown a willingness to take on a larger military role on the global stage not only to fulfill its international obligations to keep the SLOCs safe, but also to protect a more diverse set of national interests that are not defined by geography, such as protecting economic interests, preventing terrorism, delivering humanitarian aid, and conducting other military operations other than war (MOOTW).[51] Beijing's deployment of a task force made up of two guided missile destroyers and a supply ship to participate in international anti-piracy operations in the Gulf of Aden demonstrates its acceptance of its international responsibility to protect international commerce, not just its own.[52]

China seeks to expand military ties through joint military exercises and high-level exchanges at the defense minister level, regional level, military academy level, and defense industrial agency level.[53] Beijing also seeks to achieve its political objectives of regional security and stability on substantive issues of mutual concern, such as counternarcotics trafficking, smuggling, and terrorism and to expand cooperative relations with all nations in the region.[54]

In terms of energy security, China would seek to prevent hostile powers from interfering with the delivery (by land or sea) of its oil supply from the Near East and Africa.[55] Diversification strategies such as robust overland transportation systems between Yunnan and the Bay of Bengal and between Xinjiang Province in Western China and Gwadar, Pakistan—as well as military strategies such as Chinese naval escorts from Near East ports to China—would reduce China's energy insecurity in South Asia and substantially strengthen the Chinese military's ability to sustain intense military operations in the Indian Ocean region.[56] Further, if the Chinese Navy (PLAN) established bases on the Indian Ocean littoral (such as Gwadar and Kyaukphyu, Myanmar) and linked those forward bases to China by robust road and rail lines, America's ability to restrict oil shipments in the Indian Ocean region would be substantially diminished.[57] This ability would be further diminished by hard naval cooperation between China and India at the expense of the U.S.-Indian naval cooperation.

U.S.-China-India Diplomatic/Political Relations

U.S.-Indian diplomatic relations are very good even though India does not produce crude oil for the United States; both are among the world's largest democracies and are committed to political freedom protected by representative government. Their relationship is considered strategic in nature in which both countries share an interest in fighting terrorism and in creating a strategically stable Asia. In this respect, the United States recognizes India as a key to its strategic interests in the region and has sought to strengthen its relationship with India in counterterrorism and civilian nuclear energy cooperation.[58]

Sino-Indian diplomatic relations have improved gradually since 1988 despite suspicions remaining from a 1962 border conflict between India and China and continuing territorial/boundary disputes.[59] However, China's continued military and economic support for Pakistan, which includes nuclear weapons and missile-related transfers, is a major and ongoing source of friction.[60] China's support for Pakistan's Kashmir position adds further discomfort to bilateral relations. In addition, China's security relations with neighboring Myanmar and the construction of military and port facilities on the Indian Ocean is widely seen by Indian security strategists as an encirclement strategy designed to prevent or delay India's ability to challenge China's influence in South Asia.[61]

The construction of an oil pipeline connecting Russia, China, and India is unlikely to result in a diplomatic competition for oil energy resources between the United States and China because the United States does not depend on Russia, China, or India for its oil energy. An oil pipeline connecting Russia, China, and India would strengthen economic relations between Russia, China, and India but not necessarily lead to better diplomatic relations between China and India because of China's strategic relationship with Pakistan.

Hard naval cooperation between China and India at the expense of U.S.-Indian naval cooperation is unlikely to result in a diplomatic competition for oil energy resources between the United States and China. However, the United States would be compelled to deploy more naval assets in the Indian Ocean region to protect the sea lines of communication unless the

United States, China, and India made cooperative measures to protect maritime commerce. Moreover, America's ability to restrict oil shipments in the Indian Ocean region would be diminished. In this respect, America's ability to blockade Chinese oil tankers at the strategic chokepoints would be diminished because of Sino-Indian naval cooperation. Overall, hard naval cooperation between China and India is very unlikely in the near future given their past history and the political differences.

U.S.-China-Pakistan Diplomatic/Political Relations

U.S.-Pakistani diplomatic relations are generally good even though Pakistan does not produce crude oil for the United States. The United States designated Pakistan as a major non-NATO ally in 2004 in recognition of their close bilateral ties.[62] However, bilateral relations are based on strategic interests rather than a genuine partnership.[63] In this respect, the United States was able to secure its political interests through Pakistan when Pakistan received American monetary aid.[64] Essentially, the United States has been a fair-weather ally to Pakistan when it served U.S. interests. Thus, the bilateral relationship is based on realpolitik and not ideological considerations.

Sino-Pakistani diplomatic relations are robust even though Pakistan does not produce crude oil for China. Their relationship is enduring, multidimensional, and deep-rooted, underpinned by mutual trust.[65] Cooperation is strategic, as evidenced by the fact that China has been a major arms supplier to Pakistan while Pakistan has supported China on domestic issues related to sovereignty and human rights.

The construction of a seaport and naval base at Gwadar, oil refineries in Baluchistan, and an oil pipeline along the Karakorum Highway linking China and Pakistan are unlikely to result in a Sino-American diplomatic competition for oil energy resources because Pakistan does not produce crude oil for the United States. However, the United States would be compelled to step up its diplomatic efforts to ensure that its strategic interests are promoted in South Asia because of Chinese competition for regional influence. In this respect, both the United States and China are competing for influence in South Asia, especially in Pakistan.

U.S.-China-India Economic Relations

Although India does not produce crude oil for the United States, U.S.-Indian economic relations are robust as evidenced by the fact that bilateral merchandise trade in 2007 was $41.6 billion.[66] The United States is India's largest trading partner as well as India's largest investment partner with a 13 percent share.[67]

In terms of trade, the United States exports to India diagnostic or lab reagents, aircraft and parts, advanced machinery, cotton, fertilizers, ferrous waste/scrap metal, and computer hardware while India exports to the United States textiles and ready-made garments, Internet-enabled services, agricultural and related products, gems and jewelry, leather products, and chemicals.[68]

Sino-Indian economic relations are very good even though India does not produce crude oil for China. China is India's principal trading partner after the United States, according to the U.S. Central Intelligence Agency.

China exports machinery and equipment, electric and electronic products, chemical products, textiles, mineral products, base metals, plastics, rubber articles, ceramic products, glass and glassware, measuring instruments, clocks and watches, and vegetable products to India while India exports mining and industrial products, base metals, chemicals, textiles and textile articles, gems, precious metals, machinery and equipment, electric and electronic products, leather, vegetable products, footwear, and headwear to China.[69]

The construction of an oil pipeline connecting Russia, China, and India is unlikely to result in an economic competition for oil energy resources between the United States and China because the United States does not depend on Russia, China, or India for its oil energy. An oil pipeline connecting Russia, China, and India would strengthen economic relations between Russia, China, and India but not necessarily at the economic expense of the United States.

Hard naval cooperation between China and India at the expense of U.S.-Indian naval cooperation is unlikely to result in a Sino-American economic competition for oil energy resources. However, the United States would be compelled to deploy more naval assets in the Indian Ocean region to protect the sea lines of communication unless the United States, China, and India engaged in cooperative measures to protect maritime commerce in the Indian Ocean region and the Persian Gulf. Moreover, hard naval cooperation between China and India at the expense of U.S.-Indian naval cooperation would reduce America's ability to blockade Chinese oil tankers at the strategic chokepoints in the event that China attacked Taiwan because India would not necessarily be an ally to the United States. In addition, Sino-Indian naval cooperation at the expense of U.S.-Indian naval cooperation could lead to a spike in world oil prices because of India's power projection capability in the Indian Ocean region. Overall, hard naval cooperation between China and India is very unlikely in the near future given the political differences between

China and India. However, it would be problematic for U.S. defense strategy in the Indian Ocean region if the scenario came to fruition.

U.S.-China-Pakistan Economic Relations

Although Pakistan does not produce crude oil for the United States, U.S.-Pakistani economic relations are generally good.[70] The United States is Pakistan's largest export trading partner, with a 21 percent share in 2007.[71] Moreover, the United States is Pakistan's largest import trading partner after China, Saudi Arabia, the United Arab Emirates and Japan.[72]

Sino-Pakistani economic relations are very good even though Pakistan does not produce crude oil for China. China is Pakistan's largest import trading partner, with a 13.8 percent share in 2007.[73] Moreover, China is Pakistan's largest export trading partner after the United States, the United Arab Emirates, Afghanistan, and the United Kingdom.[74]

The construction of a seaport and naval base at Gwadar, oil refineries in Baluchistan, and an oil pipeline along the Karakorum Highway linking China and Pakistan is unlikely to result in a Sino-American economic competition for oil energy resources because Pakistan do not produce crude oil for the United States. However, the United States may be compelled to increase its economic aid and assistance to Pakistan to ensure that its economic interests are promoted in South Asia because of rising Chinese influence and the fact that U.S.-Pakistani ties are based on realpolitik.

U.S.-China-India Military Relations

Although India does not produce crude oil for the United States, U.S.-Indian military relations are very good as evidenced by the fact that the United States and India have shared security interests and bilateral military cooperation in defeating terrorism, violent religious extremism, preventing the spread of weapons of mass destruction and associated materials, data, and technologies and protecting the free flow of commerce via land, air and sea lanes. The United States and India have conducted joint military exercises in the Indian Ocean region in recent years.[75] Further, "the United States and India signed a ten-year defense pact outlining planned collaboration in multinational operations, expanded two-way defense trade, increasing opportunities for technology transfer and co-production, expanded collaboration related to missile-defense, and the establishment of a bilateral Defense Procurement and Production Group."[76]

Sino-Indian military relations are generally good even though India does not produce crude oil for China. Bilateral military relations have been characterized by modest high-level military-to-military visits and dialogues and unprecedented combined naval and army exercises.[77]

Hard naval cooperation between China and India at the expense of U.S.-Indian naval cooperation is unlikely to result in a military competition for oil energy resources between the United States and China. However, the United States would be compelled to deploy more naval assets in the Indian Ocean region to protect the sea lines of communication unless the United States, China and India engaged in cooperative measures to protect maritime commerce. Moreover, America's ability to restrict oil shipments in the Indian Ocean region would be diminished. In this respect, America's ability to blockade Chinese oil tankers at the strategic chokepoints would be diminished because of Sino-Indian naval cooperation. Overall, hard naval cooperation between China and India is very unlikely in the near future given their past history and the political differences.

U.S.-China-Pakistan Military Relations

Although Pakistan does not produce crude oil for the United States, U.S.-Pakistani military relations are very good.[78] The United States designated Pakistan a major non-NATO U.S. ally in June 2004.[79] Moreover, since September 2001, Pakistan has provided extensive assistance in the U.S.-led global war on terror by capturing more than 600 Al-Qaeda members and their allies.[80]

Sino-Pakistani relations are robust even though Pakistan does not produce crude oil for China. Bilateral military relations have been characterized by high-level civilian and military visits and dialogues since 1956 and technical cooperation in conventional and strategic weapons programs. China's military relationship with Pakistan's military has often been regarded by the leaders of both countries as time-tested.

The construction of a seaport and naval base at Gwadar, oil refineries in Baluchistan, and an oil pipeline along the Karakorum Highway linking China and Pakistan are unlikely to result in a Sino-American military competition for oil energy resources because Pakistan does not produce crude oil for the United States. However, the United States would be compelled to ramp up its naval presence in the Indian Ocean region to ensure that its economic interests are protected because of increased Chinese naval presence in the region. In this respect, both the United States and China have overlapping strategic interests in South Asia and the Indian Ocean region, which includes the protection of the sea lines of communication (SLOCs) that commercial tankers transit to deliver crude oil to the United States and China.

Notes

1. Congressional Research Service, Library of Congress, *Comparing Global Influence: China's and U.S. Diplomacy, Foreign Aid, Trade, and Investment in the Developing World* (Washington, D.C.: U.S. Government Printing Office), RL 34620, August 15, 2008, 77.

2. See www.princeandpanda.com/media/6d4c7d4a7bb40f5bffff8019ffffa126.pdf.

3. U.S. Senate, Committee on Foreign Relations, *The Emergence of China Throughout Asia: Security and Economic Consequences for the United States*, Hearing on June 7, 2005, 109th Congress, Session 1, 2005, 45.

4. Cole, Bernard D., *Sea Lanes and Pipelines: Energy Security in Asia* (Westport, Connecticut: Praeger Security International), 2008, 56.

5. See www.princeandpanda.com/media/6d4c7d4a7bb40f5bffff8019ffffa126.pdf.

6. Congressional Research Service, Library of Congress, *China's Foreign Policy and "Soft Power" in South America, Asia, and Africa: A Study Prepared for the Committee on Foreign Relations United States Senate* (Washington, D.C.: U.S. Government Printing Office), April 2008, 52, 55.

7. Congressional Research Service, Library of Congress, *China's Foreign Policy and "Soft Power" in South America, Asia, and Africa: A Study Prepared for the Committee on Foreign Relations United States Senate,* 52, 55.

8. Congressional Research Service, Library of Congress, *China's Foreign Policy and "Soft Power" in South America, Asia, and Africa: A Study Prepared for the Committee on Foreign Relations United States Senate,* 52, 55.

9. See www.princeandpanda.com/media/6d4c7d4a7bb40f5bffff8019ffffa126.pdf.

10. Congressional Research Service, Library of Congress, *Comparing Global Influence: China's and U.S. Diplomacy, Foreign Aid, Trade, and Investment in the Developing World* (Washington, D.C.: U.S. Government Printing Office), RL 34620, August 15, 2008, 77.

11. Congressional Research Service, Library of Congress, *Comparing Global Influence: China's and U.S. Diplomacy, Foreign Aid, Trade, and Investment in the Developing World,* 77.

12. Sutter, Robert G., *Chinese Foreign Relations: Power and Policy Since the Cold War* (Landam, Maryland: Rowman & Littlefield Publishers), 2008, 92.

13. Herberg, Mikkal, "China's Search for Energy Security: the Implications for Southeast Asia," in Evelyn Goh and Sheldon W. Simon, eds., *China, United States, and Southeast Asia: Contending Perspectives on Politics, Security, and Economics* (New York: Routledge), 2008, 79.

14. Congressional Research Service, Library of Congress, *Comparing Global Influence: China's and U.S. Diplomacy, Foreign Aid, Trade, and Investment in the Developing World* (Washington, D.C.: U.S. Government Printing Office), RL 34620, August 15, 2008, 77.

15. Cole, *Sea Lanes and Pipelines: Energy Security in Asia,* 83.

16. Johnston, Corey S., *Transnational Pipelines and Naval Expansion: Examining China's Oil Insecurities in the Indian Ocean (Master's Thesis)* (Monterey, CA: Naval Postgraduate School), June 2008, 1.

17. Lieberthal, Kenneth and Mikkal Herberg, *China's Search for Energy Security: Implications for U.S. Policy* (Seattle: National Bureau of Asian Research), Volume 17, Number 1, April 2006, 11.

18. U.S. Senate, Committee on Foreign Relations, *The Emergence of China Throughout Asia: Security and Economic Consequences for the United States*, 44–45.

19. Herberg, "China's Search for Energy Security: the Implications for Southeast Asia," 82.

20. Pehrson, Christopher J., *String of Pearls: Meeting the Challenge of China's Rising Power across the Asian Littoral.* Carlisle, PA: Strategic Studies Institute, July 2006, v.

21. Herberg, "China's Search for Energy Security: the Implications for Southeast Asia," 79.

22. Herberg, "China's Search for Energy Security: the Implications for Southeast Asia," 77.

23. Leverett, Flynt, "Resource Merchantism and the Militarization of Resource Management: Rising Asia and the Future of American Primacy," in Daniel Moran and James A. Russell, eds., *Energy Security and Global Politics: the Militarization of Resource Management* (New York: Routledge), 2009, 211.

24. Lieberthal and Herberg, *China's Search for Energy Security: Implications for U.S. Policy,* 14.

25. Lieberthal and Herberg, *China's Search for Energy Security: Implications for U.S. Policy,* 15–16.

26. See footnote 3 of Moran, Daniel and James A. Russell, "Introduction: The Militarization of Energy Security," in Daniel Moran and James A. Russell, eds., *Energy Security and Global Politics: the Militarization of Resource Management.* (New York: Routledge), 2009, 17.

27. Congressional Research Service, the Library of Congress, *China's "Soft Power" in Southeast Asia* (Washington, D.C.: U.S. Government Printing Office), RL 34310, January 4, 2008, 16–17.

28. Congressional Research Service, the Library of Congress, *China's "Soft Power" in Southeast Asia,* 17.

29. Goh, Evelyn and Sheldon W. Simon, "Introduction," in Evelyn Goh and Sheldon W. Simon, eds., *China, the United States, and Southeast Asia: Contending Perspectives on Politics, Security, and Economics* (New York: Routledge), 2008, 1.

30. Herberg, "China's Search for Energy Security: the Implications for Southeast Asia," 71.

31. Chambers, Michael, "China's Military Rise to Great Power Status: Its Implications for the United States in Southeast Asia," in Evelyn Goh and Sheldon W. Simon,

eds., *China, the United States, and Southeast Asia: Contending Perspectives on Politics, Security, and Economics* (New York: Routledge), 2008, 172.

32. U.S. Department of State, Background Notes: India, *www.state.gov*, January 2009.

33. Matheswaran, M., "Air and Aerospace Partnership" in Swaran Singh, ed., *China-Pakistan Strategic Cooperation: Indian Perspectives* (New Delhi, India: Manohar Publishers), 2007, 198.

34. Sakhuja, Vijay, "Naval Cooperation" in Swaran Singh, ed., *China-Pakistan Strategic Cooperation: Indian Perspectives* (New Delhi, India: Manohar Publishers), 2007, 217.

35. Sakhuja, "Naval Cooperation," 217–19.

36. Sakhuja, "Naval Cooperation," 220.

37. Sakhuja, "Naval Cooperation," 215–16.

38. Prabhakar, W. Lawrence, "The Maritime Convergence" in Swaran Singh, ed., *China-Pakistan Strategic Cooperation: Indian Perspectives* (New Delhi, India: Manohar Publishers), 2007, 238.

39. Johnston, *Transnational Pipelines and Naval Expansion: Examining China's Oil Insecurities in the Indian Ocean (Master's Thesis),* 83.

40. Prabhakar, "The Maritime Convergence," 238

41. Niazi, Tarique, "Gwadar: China's Naval Outpost on the Indian Ocean," *Jamestown Foundation*, February 28, 2005.

42. Lieberthal and Herberg, *China's Search for Energy Security: Implications for U.S. Policy,* 24.

43. See www.princeandpanda.com/media/6d4c7d4a7bb40f5bffff8019ffffa126.pdf.

44. Peters, John E., James Dickens, Derek Eaton, C. Christine Fair, Nina Hachigian, Theodore W. Karasik, Rollie Lal, Rachel M. Swanger, Gregory F. Treverton, and Charles W. Wolf, *War and Escalation in South Asia* (Santa Monica, California: RAND, MG-367–1), 2006, 2.

45. Garver, John W., *China's South Asian Interests and Policies*, Prepared for the panel on "China's Approaches to South Asia and the Former Soviet States," U.S.-China Economic and Security Review Commission, July 22, 2005.

46. Garver, *China's South Asian Interests and Policies.*

47. Garver, *China's South Asian Interests and Policies.*

48. U.S. Department of State, Background Notes: Pakistan, *www.state.gov,* March 2009.

49. Peters, John E., James Dickens, Derek Eaton, C. Christine Fair, Nina Hachigian, Theodore W. Karasik, Rollie Lal, Rachel M. Swanger, Gregory F. Treverton, and Charles W. Wolf, *War and Escalation in South Asia* (Santa Monica, California: RAND, MG-367–1), 2006, 2.

50. Bhalla, Madhu, "Geopolitics of Economic Relations" in Swaran Singh, ed., *China-Pakistan Strategic Cooperation: Indian Perspectives* (New Delhi, India: Manohar Publishers), 2007, 251.

51. Kaufman, Alison A., *China's Participation in Anti-Piracy Operations off the Horn of Africa: Drivers and Implications* (Alexandria, Virginia: Center for Naval Analysis), July 2009, 1, 4, 7.

52. Kaufman, *China's Participation in Anti-Piracy Operations off the Horn of Africa: Drivers and Implications,* 1.

53. Garver, *China's South Asian Interests and Policies.*

54. Garver, *China's South Asian Interests and Policies.*

55. Garver, *China's South Asian Interests and Policies.*

56. Garver, *China's South Asian Interests and Policies.*

57. Garver, *China's South Asian Interests and Policies.*

58. U.S. Department of State, Background Notes: India, *www.state.gov*, January 2009.

59. U.S. Department of State, Background Notes: India, *www.state.gov*, January 2009.

60. Alan K. Kronstadt, *India-U.S. Relations,* Congressional Research Service, the Library of Congress, RL33529, January 30, 2009, 19.

61. Kronstadt, *India-U.S. Relations,* 19.

62. U.S. Department of State, Background Notes: Pakistan, *www.state.gov,* March 2009.

63. U.S. Department of State, Background Notes: Pakistan, *www.state.gov,* March 2009.

64. U.S. Department of State, Background Notes: Pakistan, *www.state.gov,* March 2009.

65. U.S. Department of State, Background Notes: Pakistan, *www.state.gov,* March 2009.

66. U.S. Department of State, Background Notes: India, *www.state.gov*, January 2009.

67. U.S. Department of State, Background Notes: India, *www.state.gov*, January 2009.

68. U.S. Department of State, Background Notes: India, *www.state.gov*, January 2009.

69. *http://www.chinaembassy.org.in/eng/jjmy/t193966.htm.*

70. U.S. Department of State, Background Notes: Pakistan, *www.state.gov,* March 2009.

71. U.S. Department of State, Background Notes: Pakistan, *www.state.gov,* March 2009.

72. U.S. Department of State, Background Notes: Pakistan, *www.state.gov,* March 2009.

73. U.S. Department of State, Background Notes: Pakistan, *www.state.gov,* March 2009.

74. U.S. Department of State, Background Notes: Pakistan, *www.state.gov,* March 2009.

75. U.S. Department of State, Background Notes: India, *www.state.gov*, January 2009.

76. Kronstadt, *India-U.S. Relations,* 48.

77. U.S. Department of State, Background Notes: India, *www.state.gov*, January 2009.

78. U.S. Department of State, Background Notes: Pakistan, *www.state.gov,* March 2009.

79. U.S. Department of State, Background Notes: Pakistan, *www.state.gov,* March 2009 and Kronstadt, *Pakistan-U.S. Relations*, 53.

80. U.S. Department of State, Background Notes: Pakistan, *www.state.gov,* March 2009.

Chapter 4

Central Eurasia

In terms of the oil-producing states, the Central Eurasian region is comprised of Azerbaijan, Kazakhstan, Turkmenistan, and Uzbekistan. Many observers see this region as an area of vital Sino-American competition because of overlapping security interests. In terms of U.S. energy trading, crude oil and petroleum imports ranks last in the world.

The PRINCE model predicts that China will continue to explore for oil in and import oil from Central Eurasia based on the following probabilities: 97 percent Azerbaijan, 95 percent (Kazakhstan 93 percent Turkmenistan, and 91.4 percent Uzbekistan.[1]

POLITICAL, ECONOMIC, AND MILITARY IMPLICATIONS FOR THE UNITED STATES

China has four sets of interests and policies in Eurasia.[2] First, China has strategic and diplomatic goals that play into its foreign policy with respect to Sino-U.S. relations and Sino-Russian relations. Second, China seeks to safeguard the stability and security of its borders with the Central Asian nations. Third, China seeks to prevent any external influence in its western region, including support for the Uighur separatists in Xinjiang. Finally, China has energy and trade interests, which include the development of energy resources including oil and gas.

None of China's interests and policies appear to challenge U.S. interests in the region because the United States and China share similar goals such as combating terrorist activity, especially in the wake of the September 11, 2001, terrorist attacks against the United States.[3] Further, China is aware to U.S. sensitivities in the region in the post-September 11 environment and wants to avoid perceptions of a "rising China" as a regional hegemon.[4] In addition, the region's geographic position—landlocked and remote from most U.S. interests—makes it of little strategic consequence for Washington. Moreover, in terms of U.S. energy security, the region is a very minor oil producer and contains less than 4 percent of the world's proven oil reserves. Finally, China is just beginning to realize the region's importance to its energy security, especially for diversification purposes. Therefore, it is unlikely that China's energy trading with the region will result in a Sino-American diplomatic competition for oil energy resources.

Although China's oil energy stakes in Eurasia will not result in a Sino-American diplomatic competition for oil energy resources, it is likely that China's soft power approach to (and engagement in) Eurasia may hamper U.S. efforts to promote democratization, good governance and transparency reforms, universal norms of civic and human rights, and the rule of law in the region, because unlike the United States, China adheres to the principle that the domestic affairs of a country should not be subject to international interference.[5] The fact that the United States stipulates conditions for aid, including the funding of NGOs, places it at odds with China. However, Eurasian states are wary of Chinese objectives and "view the U.S. presence in the region as an insurance policy against any future bid by China and Russia to reassert control."[6] Therefore, any U.S-planned initiative for a Greater Eurasia that includes a grouping of countries friendly to American policy may serve as an effective counterweight to any regional political organization under significant Chinese influence, such as the Shanghai Cooperation Organization.

U.S. economic interests in Central Eurasia are primarily to develop the region's oil and gas reserves, which may help the region to develop some economic independence from Russia.[7] In particular, the United States has supported pipeline projects operating from the Caspian littoral states, but logistical challenges make it difficult to transport hydrocarbons over land. The Caspian region has economic potential, but the region's limited energy resources do not make it a high priority for U.S. foreign policy.

Economically, China seeks to develop its energy and trade relations with Central Eurasia. As early as 1994, China called for the construction of a new "Silk Road" to connect Central Eurasia (especially Central Asia) with China and act as a conduit for trade between Asia and the Middle East and Europe.[8] However, China's economic ambitions have yet to be realized. Nevertheless, China has made significant progress especially with regard to the development of the region's enormous energy resources to fuel its economic growth and burgeoning energy demand.[9] It should be noted that even though the

Caspian Sea is very promising in terms of energy resources, it is not yet a significant source of Chinese energy.[10]

Central Eurasia plays an important role to China's energy diversification and energy security. The region is less vulnerable to U.S. power than the Persian Gulf or the sea lanes connecting the Persian Gulf to the South China Sea.[11] Further, the United States' ability to threaten China's oil supply from the region is limited because of geography.[12] Geography makes the region's energy cooperation greater with China than the United States.

China's energy stakes in Central Eurasia are unlikely to result in a Sino-American economic competition for oil energy resources because the region is a very low producer of oil to both the United States and China. Further, the region's proven oil reserves are very low compared to other resource-rich regions of the world like Africa, the Near East, and the Western Hemisphere. Thus, Central Eurasia is inconsequential to both America's and China's energy security. Nevertheless, China will continue to develop the region for its oil energy resources because of its geographic proximity and its energy security vulnerability.

The United States seeks to bring security and stability to the Central Eurasian region in the wake of the September 11, 2001 attacks on New York and Washington. Prior to September 11, transnational threats to the peace and stability of the region—such as organized crime, terrorism, proliferation and narcotics trafficking—received relatively little attention.[13] However, the September 11 attacks clearly and unquestionably demonstrated that state failure can present threats far beyond the borders of the state in question.[14] Moreover, failed states can "create environments for groups or individuals that seek to threaten nations or populations."[15] Finally, it illustrated that dangerous links can exist between the transnational threats that can flourish in such environments. [16] Therefore, Central Eurasia has become a national security priority for the United States since September 11. Although the current U.S. priority is the eradication of terrorism, the long-term goal is to construct a new political, social, and economic environment that will discourage the emergence of security problems.[17]

According to Bates Gill and Matthew Oresman, China's primary security goal in Central Asia is prohibiting the region from becoming an external base of support for separatist Uyghur elements in China's northwestern Xinjiang province.[18] Further, China will assist the region in controlling its own domestic security challenges, including terrorism, separatism, religious extremism, and drug trafficking, so that these problems do not become a security threat to China itself.[19]

China's maintains its security in the Eurasian region principally through the Shanghai Cooperation Organization (SCO), which was founded in Shanghai in 1995 as the Shanghai Five by China, Russia, Kazakhstan, Kyrgyzstan, and Tajikistan.[20] The SCO was initially a mechanism for settling border disputes between member states and protecting the security of each nation.[21] However, since the September 11 attacks, the organization has shifted into an anti-terrorist organization.[22] Although SCO has met with mixed success in this respect, it has enabled the Central Asian states to take part in generating regional approaches to cooperation and security on an equal basis with large regional powers like China and Russia.[23]

China's energy stakes in Central Eurasia is unlikely to result in a Sino-American military competition for oil energy resources because the region is a very low producer of oil to both the U.S. and China. Further, the region's proven oil reserves are very low compared to other resource rich regions of the world like Africa, the Near East and the Western Hemisphere. Thus, Central Eurasia is inconsequential to both America and China's energy security. However, the region is strategically vital to both America and China's national security because it is a potential area of political instability including a haven for terrorists to recruit, train and operate from.

Notes

1. See www.princeandpanda.com/media/6d4c7d4a7bb40f5bffff8019ffffa126.pdf.

2. Oresman, Matthew, "Repaving the Silk Road: China's Emergence in Central Asia," in Joshua Eisenman, Eric Heginbotham, and Derek Mitchell, eds., *China and the Developing World: Beijing's Strategy for the Twenty-First Century* (Armonk, New York: M.E. Sharpe), 2007, 61–62; Bates Gill and Matthew Oresman, *China's New Journey to the West: China's Emergence in Central Asia and Implications for U.S. Interests* (Washington, D.C.: CSIS Press), August 2003, 13; and Thrassy N. Marketos, *China's Energy Geopolitics: The Shanghai Cooperation Organization and Central Asia* (New York: Routledge), 2009, 11.

3. Gill and Oresman, *China's New Journey to the West: China's Emergence in Central Asia and Implications for U.S. Interests,* 1–2.

4. Gill and Oresman, *China's New Journey to the West: China's Emergence in Central Asia and Implications for U.S. Interests.* 2.

5. Congressional Research Service, the Library of Congress, *Comparing Global Influence: China's and U.S. Diplomacy, Foreign Aid, Trade, and Investment in the Developing World* (Washington, D.C.: U.S. Government Printing Office), RL 34620, August 15, 2008, 95.

6. Marketos, *China's Energy Geopolitics: The Shanghai Cooperation Organization and Central Asia,* 107, 113.

7. Marketos, *China's Energy Geopolitics: The Shanghai Cooperation Organization and Central Asia,* 29–30.

8. Gill and Oresman, *China's New Journey to the West: China's Emergence in Central Asia and Implications for U.S. Interests,* 22.

9. Gill and Oresman, *China's New Journey to the West: China's Emergence in Central Asia and Implications for U.S. Interests,* 22.

10. Gill and Oresman, *China's New Journey to the West: China's Emergence in Central Asia and Implications for U.S. Interests,* 28.

11. Marketos, *China's Energy Geopolitics: The Shanghai Cooperation Organization and Central Asia,* 105.

12. Marketos, *China's Energy Geopolitics: The Shanghai Cooperation Organization and Central Asia,* 105.

13. Oliker, Olga and David A. Shlapak, *U.S. Interests in Central Asia: Policy Priorities and Military Roles* (Santa Monica, California: RAND), MG-338, 2005, 19.

14. Oliker and Shlapak, *U.S. Interests in Central Asia: Policy Priorities and Military Roles, 19.*

15. Oliker and Shlapak, *U.S. Interests in Central Asia: Policy Priorities and Military Roles, 19.*

16. Oliker and Shlapak, *U.S. Interests in Central Asia: Policy Priorities and Military Roles, 19.*

17. Gill and Oresman, *China's New Journey to the West: China's Emergence in Central Asia and Implications for U.S. Interests,* 39.

18. Gill and Oresman, *China's New Journey to the West: China's Emergence in Central Asia and Implications for U.S. Interests,* viii.

19. Gill and Oresman, *China's New Journey to the West: China's Emergence in Central Asia and Implications for U.S. Interests,* viii.

20. Marketos, *China's Energy Geopolitics: The Shanghai Cooperation Organization and Central Asia,* 4.

21. Marketos, *China's Energy Geopolitics: The Shanghai Cooperation Organization and Central Asia,* 32.

22. Marketos, *China's Energy Geopolitics: The Shanghai Cooperation Organization and Central Asia,* 14.

23. Marketos, *China's Energy Geopolitics: The Shanghai Cooperation Organization and Central Asia,* 33.

Chapter 5

The Middle East and North Africa (Near East)

In terms of the energy-producing states for the United States and China, the Middle East and North Africa region is made up of Algeria, Egypt, Iran, Iraq, Kuwait, Libya, Morocco, Oman, Qatar, Saudi Arabia, Syria, Tunisia, the United Arab Emirates, and Yemen. Many observers see this region as the key area of Sino-American competition because both China and the United States have overlapping diplomatic, commercial, and security interests in the region. In terms of U.S. energy trading, crude oil and petroleum imports from the Middle East and North Africa rank second after the Western Hemisphere, with Canada and Mexico together being America's largest hydrocarbon suppliers. Comparatively, China has imported more crude oil from the Middle East and North Africa than from any other region in the world. Almost 50 percent of China's crude oil and petroleum is imported from the Middle East and North Africa.

While it appears that China's interest in the Near East is mainly economic, nevertheless it has adopted a comprehensive, three-pronged strategy to promote its overall interests. First, China uses diplomacy to help Chinese businesses make inroads to the oil industry and markets of the Middle East and North Africa.[1] Second, China has "adopted an offend-no-one approach to regional politics in order to maintain an environment friendly to Chinese business interests."[2] Third, China exploits its soft-power resources to cultivate goodwill and friendship with an eye toward protecting its long-term economic interests.[3] China's three-pronged strategy as well as strained relations between the Middle East and United States in the wake of September 11, 2001 and the U.S-led war in Iraq and Afghanistan has made China a much more attractive strategic partner for the Middle East to the dismay of the United States, which creates possibilities for a future Sino-American conflict. Thus, China's rising concentration on the Middle East as an energy resource, and the enduring U.S. concentration on the region as a key strategic battleground creates the possibility of competition or even military conflict. [4]

The PRINCE model predicts that China will continue to explore for oil in and import oil from the Near East based on a probability of 90.4 percent or greater.[5]

POLITICAL IMPLICATIONS FOR THE UNITED STATES

"The binding cord in the triangular relationships between China, the Middle East and the United States is energy,"[6] even though China and the United States have deep political/diplomatic, commercial, and security ties in the region. Both China and the United States import considerable quantities of crude oil from the Middle East and rely on strong political/diplomatic relations with their oil suppliers to ensure that their energy lifelines are not severed. Further, "Chinese dependence on oil from this region is so extensive that any supply disruption could have significant negative effects on the Chinese economy."[7]

In terms of diplomatic relations, the United States has developed ties with Egypt, Kuwait. and the United Arab Emirates. whereas China has cultivated similar relationships with Algeria, Iran, Iraq, Oman, and Syria.[8] Only Saudi Arabia has managed to develop a balanced relationship between China and the United States, largely because of its decisive role in global energy markets.[9] However, the United States bears the "scars of more than a half century of efforts to impose both security and order of anti-imperialist sentiment that has existed before the United States entered the region and with the Cold War rhetoric that was sustained for almost half a century" despite its long, deep, and often strained relationships with some Middle Eastern States. [10] Further, the United States has sought to democratize the Middle East and North Africa because the region has been a major source of political instability and international terrorism. The Bush administration attributed the root of terrorism to authoritarian governments in Middle Eastern countries. Therefore, democratizing them became its policy goal.[11] The Middle Eastern monarchies and other authoritarian regimes, moderate and radical, felt threatened by the Bush administration's democracy plan.[12] In contrast, "China has more recent, less involved friendships with essentially *all* Middle East states."[13] China has been more adept at managing its foreign policy contradictions in the sense that China has clear and limited foreign policy priorities and is better able to act according to them vis-à-vis the United States, such as selling arms to antagonists on both sides on a number of disputes: Iranians and Iraqis in the 1980s, Israelis and Arabs in the 1990s, and Saudis and Iranians in the 2000s.[14] Alterman and Garver opined that "China has been able to increase its influence in the Middle East precisely by saying that it opposes any outside nation's influence and become more powerful by asserting that it distrusts power."[15]

In light of the fact that China imports more hydrocarbons than the United States on a proportional scale from essentially the same suppliers, that China appears to be more recently politically/diplomatically attractive because it does not impose conditions and lacks the "West's unfavorable history and image in the Middle East."[16] and that the United States-Middle East relations have been considerably strained since September 11, 2001 and the subsequent U.S.-led global war on terror, it is very likely that China will play a greater role in the Middle East in the near future to the detriment of the American foreign policy. In this respect, China could be a potential mediator of the Arab-Israeli conflict. China has already conducted high-level meetings with Israeli and Palestinian officials in Beijing in 2006 to the shock of Western observers, whereby "all three parties agreed that China should play a larger role in the Middle East peace process."[17] Moreover, China could potentially disrupt U.S. efforts to sanction rogue states like Iran if Iran continues on its path of nuclear weapons development given that China has considerable hydrocarbon energy stakes in Iran: $70 billion in exploration and development projects and more than 12 percent of crude oil imports. China is also known to be a source of WMD- and missile-related technologies to Iran according to the U.S.-China Economic and Security Review Commission's 2005 annual report to Congress. Therefore, China's extensive hydrocarbon energy trading and investments in the Middle East and North Africa will make it not only a potential energy competitor to the United States but also a political/diplomatic one, in the wake of China's growing political/diplomatic ties and popularity in the region, unless the United States can change its public image from that of a regional hegemon to a regional partner that will respect the politics and sovereignty of the individual countries of the Middle East and North Africa. However, at the present time, China is geostrategically excluded from the Middle East, with the possible exception of Iran.[18]

UNITED STATES-CHINA-IRAN-DIPLOMATIC/POLITICAL RELATIONS

American-Iranian diplomatic relations are cold, especially in light of Iran's historic support and involvement of international terrorism, its efforts to acquire nuclear weapons and other weapons of mass destruction, its support for violent opposition to the Middle East peace process, its dismal human rights record and lack of respect for its own people, and its actions against non-belligerent shipping in the Persian Gulf. Iran is the most contentious political issue in the Middle East vis-à-vis China.

The Congressional Research Service opined that "Iran is now perhaps the most significant political sticking point between the United States and China in the Middle East, as Chinese commercial interests have clashed with U.S. efforts to isolate Iran internationally and prevent further development of its uranium technology."[19]

In terms of isolating Iran, the U.S. has imposed international sanctions, which have impacted the development of Iran's petroleum industry. However, the sanctions have been unable to stop China's involvement in Iran's petroleum development. Sinopec is currently developing the Yadavaran oil field in southwestern Iran. In this respect, the Congressional Research Service opined that:

> Chinese investment in Iran's energy industry is the most significant example of this trend, as many in Congress and the (Bush) Administration believe that investment in Iran's energy sector could provide the Iranian government with additional revenue generating ability that would limit the effectiveness of international financial sanctions.[20]

Another factor thwarting America's efforts to impose sanctions on Iran is Iran. Iran has been playing the energy card with China to influence international policy at the United Nations. Iran has been using China's permanent member status on the United Nations Security Council to veto or threaten to veto potential new resolutions, thereby making China a powerful asset of Iran.[21] While China has voted in favor of a number of United Nations Security Council resolutions sanctioning Iran, it has also consistently worked to water them down.[22] To date, China's lack of decisive support on Iran's nuclear proliferation issue has blocked international consensus for imposing tougher sanctions on Iran.[23]

Overall, with respect to Iran, China's continued hydrocarbon trade and ongoing petroleum development projects, as well as its lack of diplomatic will to impose tougher sanctions, sets the stage for future confrontation with the United States on a grand diplomatic scale. The strategic implications for the United States vis-à-vis China are very profound given China's rising diplomatic influence in the Middle East and North Africa. It would behoove China to reconsider and appropriately calibrate its diplomatic relationship vis-à-vis Iran given the potentially dangerous situation that China is in with respect to the United States.

UNITED STATES-CHINA-IRAQ DIPLOMATIC/POLITICAL RELATIONS

Strong political/diplomatic ties define the U.S.-Iraqi relationship even though Iraq supplies more than 5 percent of America's annual crude oil. Therefore, it is unlikely that China's current energy ties with Iraq, such as crude oil trading or the development of the al-Ahdab field, will shape or degrade either current or near future U.S.-Iraqi diplomatic relations. The United States is more likely to shape Iraq's future foreign relations, since it militarily defeated Iraq and is managing Iraq's political transformation.

The United States is also more likely to shape Iraq's energy sector, including the petroleum industry, through U.S. oil companies based on ongoing negotiations by Chevron, Shell, and ExxonMobil to develop several Iraqi oil fields.[24] In this respect, "Iraq possesses the world's fourth-largest proven petroleum reserves, but many of its fields are significantly underdeveloped even though extraction costs are relatively low."[25] Iraq needs high investments in its energy infrastructure to raise production capacities, which can be provided mostly by Western oil companies because Western oil companies have the required experience, technical knowledge, and experience vis-à-vis China's nationally-owned oil companies, CNPC, Sinopec, and CNOOC. Further, security constraints on Iraqi's energy production and transportation infrastructure have severely limited exports to Asia markets, including China.[26] Since the United States is currently providing security for Iraq's energy production and transportation infrastructure through the highly capable U.S. military, it is highly unlikely that China will have any political influence over Iraq's energy sector for the time being until Iraq can break free from America's hegemonic grip.

UNITED STATES-CHINA-SAUDI ARABIA DIPLOMATIC/ POLITICAL RELATIONS

Although Saudi Arabia is America's third largest supplier of petroleum (after Canada and Mexico), accounting for almost 13 percent of America's annual crude oil imports, and has been one of America's closest allies in the Middle East and North Africa since 1938, when crude oil was first discovered in Saudi Arabia, diplomatic relations were strained after the September 11 terrorist attacks because 15 of the 19 hijackers/suicide bombers were Saudi citizens, and Saudi Arabia was publically perceived as soft on international terrorism. However, diplomatic relations have improved somewhat after May 2003 because Saudi Arabia increased its counter-terrorism efforts and cooperation with its Western law enforcement counterparts.

Economics such as hydrocarbon energy trading and investment activities as opposed to politics such as international terrorism form the crux of the Sino-Saudi relationship. Saudi Arabia is China's largest source of crude oil, accounting for 18 percent of China's annual crude oil imports. Moreover, China is Saudi Arabia's most important market for its 'distressed' medium-grade oil-a viscous, acidic, often sulfurous product that Saudi Arabia has in abundance but that has few takers on the international market. [27] Consequently, Saudi Arabia has been working to develop China's capacity to develop and use Saudi heavy crude, investing in two refineries along China's coast: the upgrading of the Quanzhou refinery in Fujian Province and the development of the Qingdao refinery in Shandong province.[28]

A robust economic relationship has formed between China and Saudi Arabia, given that the former is the world's fastest-growing energy market and that the latter is the world's largest resource holder, diplomatic ties appear to be developing between the two countries, which portend trouble for the United States. Since September 2004, a strategic alliance has been growing between China and Saudi Arabia as a political counterweight to American influence in the Middle East and North Africa as early as September 2004 because both China and Saudi Arabia oppose U.S. efforts to democratize the Middle East and North Africa by means of the Broader Middle East and North Africa Initiative (BMENA).[29]

Alterman and Garver opined, "Both capitals saw BMENA as unwarranted interference in the domestic politics of other countries, as U.S. ethnocentric arrogance, and worst of all, as an incitement to greater extremism and instability in the region."[30]

Consequently, China and Saudi Arabia signed an agreement arranging for regular political consultations. Further, Chinese President Hu Jintao and Saudi Arabian King Abdallah conducted high-level exchange visits to each other's countries in early 2006, in which President Hu laid a four-point proposal for enhancing Sino-Saudi relations. One point called for the mutual safeguarding of each other's national sovereignty and territorial integrity and the continued support and cooperation on international and regional affairs.[31] This statement implicitly targeted American efforts to democratize the Middle East and North Africa through BMENA.[32] Thus, "underlying the new Sino-Saudi political relationship is opposition to Western and especially U.S. insistence on global political norms" because both China and Saudi Arabia feel threatened by the United States.[33]

In summary, the growing Sino-Saudi political/diplomatic relationship demonstrates that China may develop and use its relationships with other

Middle Eastern countries to help nullify what it views as U.S. hegemonic aspirations.[34] Both China and the United States must act responsibly to mitigate any further damage not only in their diplomatic relations with the Middle East and North Africa but with each other.

ECONOMIC IMPLICATIONS FOR THE UNITED STATES

China's interest in the Middle East and North Africa is primarily economic: energy trading, energy cross-investment activities, and the international trade of goods and services. In terms of energy trading, the Middle East and North Africa is China's major crude oil supplier. Moreover, in terms of cross-investment activities, the Middle East and North Africa has provided China with upstream, downstream, and infrastructure development opportunities in Algeria, Egypt, Iran, Iraq, Libya, Oman, Qatar, Syria, Tunisia, and Yemen. Further, China has sought capital from the Middle East and North Africa to invest in the development of China's oil industry. Finally, in terms of goods and services, the Middle East and North Africa is a major import market for lower- and intermediate-end consumer products, including machinery and electronic products, clothes, textiles, shoes, tires, and handbags.[35]

Although China depends on the Middle East and North Africa for almost 50 percent of its crude oil, it is unlikely that China's crude oil energy trading with and cross-investment activities in the Middle East and North Africa will shift the balance of economic power from the United States to China in the near future, not only because of America's strong economic ties to the region but also because of China's inability to project naval power beyond the Pacific Ocean to protect the vital sea lines of communication for which the Near East is so heavily dependent on for its international trade. In this respect, the United States is the only power in the world that can ensure the safe and timely flow of crude oil by ocean tanker originating from the Middle East and North Africa, for which both China and the Middle East and North Africa are economically dependent.

With the fastest growing economy in the world, it is possible that China's high demand for crude oil from the Middle East and North Africa could contribute to a rise in international oil prices, which may affect the U.S. economy. In this respect, the Organization of Petroleum Exporting Countries (OPEC) could use China's high demand for crude oil as an excuse to cut oil production and raise world oil prices—which, in turn, may have dire economic consequences for the United States. Beyond OPEC, the Organization of Arab Petroleum Exporting Countries (OAPEC), a regional, multi-governmental organization that coordinates the energy policies in 11 Arab nations, may use its ability as a regional organization to influence OPEC to either raise world oil prices or embargo any crude oil destined for the United States in times of regional political crisis such as OPEC has done in the past with the Yom Kippur War, the Iranian Revolution, and the 1990 Gulf War. However, it is important to note that the International Energy Agency (IEA), a Paris-based intergovernmental organization founded by the Organization for Economic Co-operation and Development (OECD) in 1974, was created initially to prevent disruptions in the supply of oil in the wake of the 1973 oil crisis. Recently, the IEA expanded its mandate to include energy security and economic development. Therefore, the IEA could be an effective economic counterweight to any actions taken by either OPEC or OAPEC.

It is possible that China's burgeoning energy ties with the Middle East and North Africa could also lead to greater non-energy trade relations to the detriment of the United States. In this respect, China could use its energy diplomacy and trade relations to further its military arms sales to the Middle East and North Africa. China has already supplied more than $1.2 billion in military equipment to Algeria, Iran, Kuwait, and Libya between 1995 and 2002.[36] Further, China sold Saudi Arabia 50 CSS-2 Intermediate range ballistic missiles, which are capable of carrying a nuclear payload.[37] While Chinese-made weapons are generally inferior to American-made weapons, Chinese military technology could improve over time through reverse engineering to take some of the military market share away from the United States.

UNITED STATES-CHINA-IRAN-ECONOMIC RELATIONS

Although China has robust hydrocarbon energy interests in Iran such as major crude oil imports from Iran and ongoing upstream sector activities in Iranian oil fields, it is unlikely that Chinese energy activities will have either short-term or long-term economic consequences for the United States unless Iran uses China's high demand for crude oil as an excuse to cut oil production and raise oil prices. However, Iran has the potential to disrupt the flow of commercial oil transiting through the Persian Gulf, which would have dire economic consequences for the United States in light of the fact that the United States imports almost 19 percent of its crude oil annually from Iraq, Kuwait, Oman, Saudi Arabia, and the United Arab Emirates. Iran could attack commercial oil tankers in the Persian Gulf and in the Straits of Hormuz as it has done in the past during the Tanker War (1984–1988), and the U.S. Navy would be called out to respond militarily. Further, this scenario could escalate into a military conflict that would affect not only the economy of the Persian Gulf states but also the economy of China, given that China depends on the region for almost 65 percent of its crude oil.

UNITED STATES-CHINA-IRAQ ECONOMIC RELATIONS

Although China has burgeoning hydrocarbon energy interests in Iraq such as low crude oil imports from Iraq and a major equity share and upstream sector activities in Iraqi oil fields, it is unlikely that Chinese energy activities will have either short-term or long-term economic consequences for the United States unless Iraq uses China's high demand for crude oil as an excuse to cut oil production and raise oil prices.

Further, the United States is carrying out a multibillion-dollar program to assist in the reconstruction of Iraq.[38] Moreover, U.S. oil companies are developing Iraqi's oil industry in the wake of the Iraq War. Finally, security constraints on Iraqi's energy production and transportation infrastructure have severely limited exports to Asian markets, including China.[39] Therefore, it is unlikely that China's energy activities in Iraq will shape or influence U.S.-Iraqi economic relations or will have economic consequences for the United States in light of the fact the United States has a high stake in Iraqi's economic future vis-à-vis China.

UNITED STATES-CHINA-SAUDI ARABIA ECONOMIC RELATIONS

China's hydrocarbon energy interests in Saudi Arabia, which include strong crude oil trading and Saudi-Arabian oil energy investment opportunities in China, could have economic consequences for the United States because of its developing political ties with Saudi Arabia. China and Saudi Arabia could form a block to oppose any form of U.S. influence or "meddling" in the Near East. Moreover, Saudi Arabia could harm the U.S. economy if it chooses to do so based on its role as the lead producer and exporter of petroleum in the Near East, the key member of OPEC and OAPEC, and the highest known source of proven oil reserves globally, which makes it not only the decisive economic actor in the region but also in the world with respect to oil energy. However, the United States is Saudi Arabia's largest trading partner, and Saudi Arabia is the largest United States export market in the Middle East.

Further, Saudi Arabia has been able to develop a bilateral relationship with both China and the United States, "largely due to its decisive role in global oil markets."[40]

MILITARY IMPLICATIONS FOR THE UNITED STATES

China's oil energy trading and activities in the Middle East and North Africa are unlikely to shift the balance of military power and influence from the United States to China or lead to a military competition for strategic energy resources because of the strong military relationships that exist between the United States and the countries of the Near East with the exception of Iran, Libya, and Syria. The Congressional Research Service stated that "the United States remains the dominant external political and military actor in the Middle East."[41]

Although China's military presence and influence in the Middle East and North Africa is limited compared to the United States, it is possible that China may use its commercial energy ties to bolster its military relations in the region as China has done with its political relations. In this respect, the Congressional Research Service stated that "as in other regions, growing commercial ties facilitated the development of closer political relationships between China and many of its Middle Eastern counterparts."[42]

China may also use its energy diplomacy and relationships to bolster its arms relationships with the Middle East and North Africa. In this respect, "no Middle Eastern state has opposed China's entrance in the weapons sales area."[43] Further, Alterman and Garver stated that "states as diverse as Israel and Iran have sought to develop arms relationships with China, driven by the direct benefits to their militaries as well as indirect political gains."[44]

If China is able to bolster its military relations and weapons sales to the Middle East and North Africa including Iran, it "has the potential to compete militarily with the United States and field disruptive technologies that could over time off set traditional U.S. military advantages absent U.S. counter-strategies."[45]

Further, China's actions could undermine international efforts to bring stability in the Middle East and North Africa in light of the fact that it views "their efforts as one of maintaining cooperative relations with all countries in the Middle East and North Africa."[46] Alterman and Garver opined that "the United States views such efforts as undermining international efforts to isolate troubling regimes and thus compel a change in their behavior."[47]

Moreover, in light of the fact that China is heavily dependent on crude oil imports to sustain its economy, it may use its weapons sales to the Middle

East and North Africa to offset the cost of imported oil from the region. China may engage in a weapon-for-oil exchange to not only strengthen its overall ties but also reduce the cost of its imported crude oil. Erica Strecker Downs opined that "militarily, it is possible that China could use its arms sales to the Middle East to foster ties to oil-producing nations and possibly to decrease its oil import bill."[48]

However, Alterman and Garver stated that "the U.S. role at this point dwarfs China's in supplying the Gulf States, although there remains potential for a greater Chinese role in the future."[49] Therefore, further analysis of China's military relations with respect to its energy relations in the Near East is warranted. It should be noted that China has defense attaches in Algeria, Egypt, Iran, Iraq, Israel, Jordan, Kuwait, Libya, Morocco, Saudi Arabia, Syria, the United Arab Emirates, and Yemen. Of these, only four do not produce oil for China: Israel, Jordan, Morocco and Syria.

UNITED STATES-CHINA-IRAN-MILITARY RELATIONS

U.S.-Iranian military relations are non-existent. The United States severed diplomatic relations with Iran on April 7, 1980.

Sino-Iranian military relations are strong and are driven primarily by China's dependence on crude oil to sustain its economy and Iran's need for a powerful ally in the Middle East as well as a weapons supplier. China is Iran's largest supplier of conventional weapons and plays a major role in Iran's military modernization. China has provided to Iran anti-ship cruise missiles, surface-to-air missiles, combat aircraft, and fast attack patrol vessels, as well as advanced technology designed to expand the versatility of Iran's burgeoning cruise missile arsenal, which have "contributed significantly to what has become the central element of Iran's military rearmament—a revitalization of its naval forces."[50] Consequently, "U.S. intelligence agencies now estimate that Iran has the ability to shut off the flow of oil from the Persian Gulf for brief periods of time, even with a Western military presence in the region."[51]

China is also Iran's principal supplier of WMD technology and assistance, including nuclear weapons, according to the Pentagon.[52] Preliminary liaisons began in the mid- to late-1980s, and both countries are known to have signed nuclear accords in 1989 and 1991, paving the way for what would become a vibrant and multifaceted nuclear partnership.

China's military technology dealings with Iran complicate U.S. Near East foreign policy efforts to bring peace and stability to the region. A recent CRS report stated that "China's military cooperation with Iran has also proven to be a recurring problem in U.S.-China relations."[53] China's hesitance to sanction Iran in light of Iran's resistance to discontinue WMD development is an ongoing political crisis that invites a future military confrontation between Iran and United States unless China decides to act prudently and take steps to de-militarize Iran of its nuclear technology and development.

UNITED STATES-CHINA-IRAQ MILITARY RELATIONS

U.S.-Iraqi military relations are rapidly developing in the wake of the Iraqi War. U.S. military forces remain in Iraq as part of the Multi-National Force-Iraq to assist the Iraqi Government in training its security forces, as well as to work in partnership with the Iraqi Government to combat forces that seek to derail Iraq's progression toward full democracy.[54]

Sino-Iraqi military relations are resuming in the wake of the Iraqi War even though Iraq accounts for less than 1 percent of China's total crude oil imports, and China has prospective petroleum development contracts in Iraq. China has conducted high-level military-to-military dialogues with Iraqi defense officials and recently agreed to sell $100 million in light weaponry to Iraq. However, these efforts have yet to translate into a strategic Sino-Iraqi military partnership that the United States would take exception to because of America's strategic relationship with Iraq.

UNITED STATES-CHINA-SAUDI ARABIA MILITARY RELATIONS

U.S.-Saudi military ties are strong despite a row in their diplomatic relations that occurred immediately after September 11, 2001. The United States and Saudi Arabia have a longstanding security relationship, which includes U.S. military training, security support, and weapons sales to Saudi Arabia. In terms of weapon sales, the United States has sold Saudi Arabia military aircraft (F-15s, AWACS, and UH-60 Blackhawks), air defense weaponry (Patriot and Hawk missiles), armored vehicles (M1A2 Abrams tanks and M-2 Bradley infantry fighting vehicles), and other equipment.[55] Moreover, the U.S. Army Corps of Engineers has had a long-term role in military and civilian construction activities in Saudi Arabia.[56]

Sino-Saudi military relations are developing in the wake of robust energy ties. China maintains a permanent defense attaché in the capital city of Riyadh and has conducted high-level talks with Saudi defense ministers. Further, China sold and transferred 50 Chinese intermediate range ballistic missiles to Saudi Arabia in 1988. Apart from these actions, the extent of the Sino-Saudi military relationship is unknown. It appears that Sino-Saudi military ties are not as extensive as the U.S.-Saudi military ties in light of the fact that the U.S. trains Saudi military personnel and supplies the Saudi military with advanced weaponry.

Notes

1. Mao, Yufeng, "China's Interests and Strategy in the Middle East and the Arab World," in Joshua Eisenman, Eric Heginbotham and Derek Mitchell, eds., *China and the Developing World: Beijing's Strategy for the Twenty-First Century*, (Armonk, New York: M.E. Sharpe), 2007, 119.

2. Mao, Yufeng, "China's Interests and Strategy in the Middle East and the Arab World," 119.

3. Mao, Yufeng, "China's Interests and Strategy in the Middle East and the Arab World," 119.

4. Alterman, Jon. B. and John W. Garver, *The Vital Triangle: China, The United States and the Middle East* (Washington, D.C.: CSIS Press), 2008, 7.

5. See http://www.princeandpanda.com/media/6d4c7d4a7bb40f5bffff8019ffffa126.pdf.

6. Alterman, Jon. B. and John W. Garver, *The Vital Triangle: China, The United States and the Middle East,* 4.

7. Mao, Yufeng, "China's Interests and Strategy in the Middle East and the Arab World," 117.

8. Alterman, Jon. B. and John W. Garver, *The Vital Triangle: China, The United States and the Middle East,* 107.

9. Alterman, Jon. B. and John W. Garver, *The Vital Triangle: China, The United States and the Middle East,* 107.

10. Alterman, Jon. B. and John W. Garver, *The Vital Triangle: China, The United States and the Middle East,* 112.

11. Jin, Liangxiang, "Sino-Arab Relations: New Developments and Trends, *Middle East Policy*, Volume XI, No. 4, Winter 2004, 115.

12. Jin, Liangxiang, "Sino-Arab Relations: New Developments and Trends, *Middle East Policy*, Volume XI, No. 4, Winter 2004, 115.

13. Alterman, Jon. B. and John W. Garver, *The Vital Triangle: China, The United States and the Middle East,* 112.

14. Alterman, Jon. B. and John W. Garver, *The Vital Triangle: China, The United States and the Middle East,* 112.

15. Alterman, Jon. B. and John W. Garver, *The Vital Triangle: China, The United States and the Middle East,* 112.

16. Alterman, Jon. B. and John W. Garver, *The Vital Triangle: China, The United States and the Middle East,* 99.

17. Alterman, Jon. B. and John W. Garver, *The Vital Triangle: China, The United States and the Middle East,* 113.

18. Davies, Martyn J, "Special Economic Zones: China's Developmental Model Comes to Africa" in Robert I. Rotberg, ed., *China Into Africa: Trade, Aid and Influence (*Washington, D.C.: Brookings Institution Press), 2008, 151.

19. Congressional Research Service, the Library of Congress, *Comparing Global Influence: China's and U.S. Diplomacy, Foreign Aid, Trade, and Investment in the Developing World* (Washington, D.C.: U.S. Government Printing Office), RL 34620, August 15, 2008, 149.

20. Congressional Research Service, the Library of Congress, *Comparing Global Influence: China's and U.S. Diplomacy, Foreign Aid, Trade, and Investment in the Developing World*, 149.

21. Alterman, Jon. B. and John W. Garver, *The Vital Triangle: China, The United States and the Middle East,* 80.

22. Alterman, Jon. B. and John W. Garver, *The Vital Triangle: China, The United States and the Middle East,* 80.

23. Hoge, Warren, "Security Council Is Stalled Over Iran's Nuclear Program," *New York Times*, March 22, 2006, *http://www.nytimes.com/2006/03/22/inter-national/middleeast/22iran.html?_r=2&pagewanted=print&oref=slogin.*

24. Alterman, Jon. B. and John W. Garver, *The Vital Triangle: China, The United States and the Middle East,* 70.

25. Alterman, Jon. B. and John W. Garver, *The Vital Triangle: China, The United States and the Middle East,* 69–70.

26. Alterman, Jon. B. and John W. Garver, *The Vital Triangle: China, The United States and the Middle East,* 70.

27. Alterman, Jon. B. and John W. Garver, *The Vital Triangle: China, The United States and the Middle East,* 58.

28. Alterman, Jon. B. and John W. Garver, *The Vital Triangle: China, The United States and the Middle East,* 59.

29. Alterman, Jon. B. and John W. Garver, *The Vital Triangle: China, The United States and the Middle East,* 34–35.

30. Alterman, Jon. B. and John W. Garver, *The Vital Triangle: China, The United States and the Middle East,* 35.

31. Alterman, Jon. B. and John W. Garver, *The Vital Triangle: China, The United States and the Middle East,* 35.

32. Alterman, Jon. B. and John W. Garver, *The Vital Triangle: China, The United States and the Middle East,* 35.

33. Alterman, Jon. B. and John W. Garver, *The Vital Triangle: China, The United States and the Middle East,* 35.

34. Alterman, Jon. B. and John W. Garver, *The Vital Triangle: China, The United States and the Middle East,* 36.

35. Yufeng Mao, "China's Interests and Strategy in the Middle East and the Arab World," 118.

36. Alterman, Jon. B. and John W. Garver, *The Vital Triangle: China, The United States and the Middle East,* 65.

37. Alterman, Jon. B. and John W. Garver, *The Vital Triangle: China, The United States and the Middle East,* 66.

38. U.S. Department of State, Background Notes: Iraq, *www.state.gov*, February 2008.

39. Alterman, Jon. B. and John W. Garver, *The Vital Triangle: China, The United States and the Middle East,* 70.

40. Alterman, Jon. B. and John W. Garver, *The Vital Triangle: China, The United States and the Middle East,* 107.

41. Congressional Research Service, Library of Congress, *Comparing Global Influence: China's and U.S. Diplomacy, Foreign Aid, Trade, and Investment in the Developing World*, 140.

42. Congressional Research Service, Library of Congress, *Comparing Global Influence: China's and U.S. Diplomacy, Foreign Aid, Trade, and Investment in the Developing World*, 140.

43. Alterman, Jon. B. and John W. Garver, *The Vital Triangle: China, The United States and the Middle East,* 66.

44. Alterman, Jon. B. and John W. Garver, *The Vital Triangle: China, The United States and the Middle East,* 66.

45. Alterman, Jon. B. and John W. Garver, *The Vital Triangle: China, The United States and the Middle East,* 96.

46. Alterman, Jon. B. and John W. Garver, *The Vital Triangle: China, The United States and the Middle East,* 97.

47. Alterman, Jon. B. and John W. Garver, *The Vital Triangle: China, The United States and the Middle East,* 97.

48. Erica Strecker Downs, *China's Quest for Energy Security* (Santa Monica, California: RAND), MR-1244-AF, 2000, 49.

49. Alterman, Jon. B. and John W. Garver, *The Vital Triangle: China, The United States and the Middle East,* 73.

50. Ilan Berman, "The Impact of Sino-Iranian Strategic Partnership" (Testimony prepared for the U.S.-China Economic & Security Review Commission), September 14, 2006, 18.

51. Ilan Berman, "The Impact of Sino-Iranian Strategic Partnership" (Testimony prepared for the U.S.-China Economic & Security Review Commission), September 14, 2006, 18.

52. Office of the Secretary of Defense, *Proliferation: Threat and Response 1996* (Washington, D.C.: U.S. Department of Defense, 1996), 14.

53. Congressional Research Service, the Library of Congress, *Comparing Global Influence: China's and U.S. Diplomacy, Foreign Aid, Trade, and Investment in the Developing World*, 149.

54. U.S. Department of State, Background Notes: Iraq, *www.state.gov*, February 2008.

55. U.S. Department of State, Background Notes: Saudi Arabia, *www.state.gov*, February 2008.

56. U.S. Department of State, Background Notes: Saudi Arabia, *www.state.gov*, February 2008.

Chapter 6

Sub-Saharan Africa

In terms of real and potential energy producing states for the United States and China, Sub-Saharan Africa (hereafter "Africa") is comprised of Angola, Chad, Republic of Congo-Brazzaville, Cote d'Ivoire, Equatorial Guinea, Gabon, Guinea, Kenya, Madagascar, Mali, Mauritania, Niger, Nigeria, South Africa, and Sudan. Although Africa has been less strategically important to China and the United States in the past, many observers now see this region as an area of potential Sino-American competition because of overlapping political, commercial, and security interests.

Politically, economically, and militarily, the United States engaged (and continues to engage) Africa chiefly because of its strategic role in the global war on terror.

Politically, China seeks to increase its influence within international institutions. With respect to Africa, China has formalized its political influence through the Forum on China-Africa Cooperation (FOCAC). Established in 2000, FOCAC allows China to institutionalize its diplomatic overtures, technical training, debt relief, loans, grants, and infrastructure projects and allows both China and the African nations to speak with one voice.[1] China also formalized its political influence though the United Nations, in which it enjoys the support of many African nations.

China also seeks African support for the "one-China principle" in its efforts to isolate Taiwan from the international community, of which China enjoys the support of the vast majority of Africa.[2] In this respect, only four African nations currently formally recognize Taiwan: Burkina Faso, Gambia, San Tome and Principe, and Swaziland.

Economically, China desires to strengthen its bilateral commercial ties with Africa. In terms of imports, China imports crude oil and minerals from Africa. In terms of exports, China seeks African markets for its manufactured goods. However, commercial trade appears to be the dominant factor in Africa's overall relations with China and the United States. In this respect, the Congressional Research Service stated:

> The dominant factor driving such ties is trade. Sino-African ties are underpinned by China's prodigious demand for Africa's plentiful commodities, notably oil and unprocessed metals and minerals, to supply its rapidly growing economy, and by African demand for Chinese goods and services.[3]

Further, Robert Rotberg opined that

> China hardly wants to colonize, but it does have immense mercantilist ambitions. It ravenously seeks raw materials-petroleum, timber, ferrochrome, cobalt, platinum, copper, diamonds, and so on. Whatever primary resources are buried beneath the soils of Sub-Saharan Africa, China needs them to feed its massive industrial surge and-ultimately America's consumer demand.[4]

Commercial energy trading is probably the most important factor in Africa's overall relations with China and United States. The value of oil as a percentage of the value of all exports from Africa to China and United States has grown faster in the last eight years, according to Table 6.1 below. Such trends indicate that Africa is becoming an increasingly important source of oil for China and suggest that Sino-U.S. competition for African oil will grow in the new millennium.

In terms of actual U.S. hydrocarbon energy trading, crude oil, and petroleum imports from Africa rank fourth in the world. Comparatively, China has imported more crude oil from Africa than any other region in world except the Near East. Approximately 30 percent of China's crude oil is imported from Africa.

To achieve its political and economic objectives in Africa, China uses a "soft approach,": agricultural and technical assistance, investment and economic support, military patronage, and peacekeeping. China's soft approach

Table 6.1 Chinese and U.S. Oil Imports from Africa as a Proportion of All Imports from Africa

	2001	*2002*	*2003*	*2004*	*2005*	*2006*
People's Republic of China	56.5%	56.8%	60.5%	65.0%	69.0%	72.9%
United States	64.3%	61.0%	66.5%	69.7%	75.1%	76.4%

Source: U.S. Department of Commerce and China Customs.

refers to humanitarian, cultural, and educational projects aimed at promoting its image and influence.[5]

Agricultural and technical assistance consists of agro-processing projects and providing expertise and agricultural equipment, science and technology transfers, and technical training.

Investment and economic support is characterized by Chinese outward foreign direct investment (FDI) in infrastructure projects, grants, debt relief, and low-cost loans to African countries. It should be noted that loans are generally secured by oil. If an African nation defaults on a loan, it pays for it in crude oil and/or petroleum.

Military patronage consists of military cooperation, sales, and assistance. In particular, military cooperation reinforces the burgeoning Sino-African strategic partnership because it helps to ensure the security of China's investments in Africa's natural resources and infrastructure.[6]

Chinese peacekeeping under a UN umbrella also helps China to achieve its political and economic objectives in Africa through legitimate means. Integrating Chinese forces into United Nations efforts helps to mitigate conflict in Africa and serve as an effective way to build military-to-military relations and political capital globally as well as local credibility in the African nation in question.[7]

Overall, China's entry into Africa has the potential to intensify political, economic and military competition in the continent vis-à-vis the United States.[8]

The PRINCE model predicts that China will continue to explore for oil in and import oil from Sub-Saharan Africa based on a probability of 84 percent or greater.[9]

POLITICAL IMPLICATIONS FOR THE UNITED STATES

Although the Bush Administration repeatedly stated it did not view China's engagement in Africa as a threat to U.S. interests in the region, administration officials actively monitored China's activities in Africa,[10] especially since Africa currently supplies the United States with 14 percent of its petroleum needs and will be providing more than 25 percent of its oil imports by 2015, according to a report prepared for the U.S. National Intelligence Council.[11] Therefore, China's crude oil trading and investments, including infrastructure projects in Africa, potentially makes it a strategic energy resource competitor to the United States in the near future, especially since the value of oil as a percentage of the value of all exports from Africa to China and United States has grown faster since 2001.

Although China's strategic engagement in Africa is essentially commercial in nature, with petroleum accessibility being the centerpiece of China's African strategy, it has the potential to undermine African and U.S. efforts to implement various good governance and anti-corruption initiatives in many developing African nations.[12] In this respect, through financial lending China can permit African countries to circumvent the conditions upon which assistance from the United States and international institutions such as the World Bank and the International Monetary Fund is predicted. Angola is a classic case in point where this practice has occurred. In 2005, Angolan government officials were on the verge of accepting a substantial loan package from the IMF provided that the West observed strict monitoring of Angola's domestic situation in the wake of a 30-year-old civil war. However, in the face of a $2 billion unconditional package from China for the right of 10,000 barrels of crude oil per day, Angola rejected the IMF's offer.[13]

China's strategic engagement in Africa also makes it a political competitor to the United States. In this respect, Africa views China as a new market and a potential alternative to traditional U.S. dominance because of China's policy of internal non-interference.[14] After many decades of neglect and underdevelopment and being unable to benefit from the globalization process, many African countries have looked forward to new partners and new economic relations beyond their historic ties to Europe and the United States.[15] China fills this void because it is new, offers a fresh perspective to Africa's foreign relations, and demands nothing politically from the host country. The only exception is that any loans made from China to the host country may require collateral such as oil or the right to partially own or drill. China may also require that the host country trade with it exclusively.

In terms of strategic engagement, China created the Forum on China-Africa Cooperation (FOCAC) in Beijing in 2000 as a comprehensive effort to build mutually beneficial economic development, trade, and political relations with Africa rooted in principles of South-South Cooperation. Each FOCAC summit has produced a concrete action plan for Sino-African cooperation.

The 2000 summit held in Beijing passed the Beijing Declaration of the Forum on China-Africa Cooperation and the Program for China-Africa Cooperation in Economic and Social Development. [16]

The 2003 meeting passed the Addis Ababa Action Plan (2004–2006) and consolidated trade relations, with China announcing the Special Preferential Tariff Treatment program, which removed import tariffs on 190 items from 25 African countries.[17]

The 2006 summit adopted the Beijing Action Plan (BAP), which covered 2007 to the end of 2009. It was reportedly the largest international event ever held in China, drawing China's top leaders and 48 high-level African government delegations, including 41 heads of state.[18] Further, at the summit, President Hu Jintao made eight major new Chinese efforts to strengthen the Sino-African strategic partnership under FOCAC.[19] China's pragmatic policy focus on economic issues through increased political/diplomatic activity has laid the foundation for the market entry of Chinese state-owned, as well as private, companies into Africa's economies including China's oil companies CNPC, Sinopec and CNOOC.[20]

China has also reached out diplomatically/politically by maintaining an extensive network of diplomats in Africa.[21] China has embassies in all but four African countries. Further, frequent high-level visits by the Chinese president and premier have bolstered China's diplomatic presence, and reciprocal exchange visits by Chinese and African VIPs have resulted in bilateral cooperation agreements and often PRC state-financed business deals.[22]

UNITED STATES-CHINA-ANGOLA DIPLOMATIC/POLITICAL RELATIONS

U.S.-Angolan diplomatic relations are developing even though Angola is America's largest supplier of petroleum in Africa after Nigeria at approximately 4 percent annually and has historically exported more petroleum to the United States than has China. Diplomatically/politically, the United States and Angola established formal relations in 1993 whereby the former sought to consolidate peace and security, promote economic prosperity, improve health, and encourage Angola's transition to democracy and respect for human rights in the wake of a long-running civil war.[23] The United States has also worked with Angola to remove thousands of land mines and help war refugees and internally displaced people return to their homes.[24]

Sino-Angolan diplomatic relations have improved since the end of the Cold War and are based on a emerging energy trading relationship as evidenced by the fact that Angola is China's largest supplier of petroleum in Africa at approximately 18 percent annually, that China offered Angola a $9 billion loan for infrastructure improvements in return for petroleum, and that China has burgeoning projects such as offshore oil prospecting and exploration and refinery development. Further, Angola offers China a lighter or "sweeter" crude oil to import, unlike Saudi Arabia's heavier or "sour" crude oil, which is more difficult to refine in terms of time and money.

In terms of oil energy competition between the United States and China, the United States wins. Competition between the world's major oil companies operating in Angola is intense, especially in light of the fact that "Angola's role in the world's oil market is significant since there are few countries in which oil reserve estimates are increasing rapidly and that produce light oil."[25] Despite this intense competition, U.S. companies like ExxonMobil and Chevron have major investments in Angola and have historical relationships with the government of Jose Eduardo dos Santos.

Although U.S. oil companies are well-represented in Angola, they must abide by the Foreign Corrupt Practices Act, unlike Chinese oil companies. Given that Chinese firms do not conform to the Foreign Corrupt Practices Act and that the proceeds from Angola's international oil sales are largely used to finance domestic improvements and infrastructure projects in Angola, China has the ability to gain access to moderate amounts of Angolan oil by offering significant amounts of low-interest financial credit to Angola without any preconditions. In this respect, China not only undermines U.S. efforts to obtain petroleum through American oil companies but also the efforts of multilateral organizations, such as the International Monetary Fund and the World Bank, to promote political reforms.[26] Moreover, China acquires some political leverage over Angola at the expense of the United States and the West.

Essentially, China's approach to securing hydrocarbon energy resources in Angola—that is, making low-interest loans in return for oil—weakens Angola's motivation to pursue democratization, good governance, and transparency reforms, undermines U.S. and Western efforts to promote democracy in Angola, which is historically known for political corruption and human rights abuses, potentially weakens America's oil energy security and diplomatically/politically benefits China. China has already done so in 2006 by providing Angola with $2 billion in aid in return for 10,000 barrels per day, thereby undercutting IMF efforts to observe a UN-mandated cease-fire monitoring.

In summary, China's crude oil and petroleum energy trading with Angola may have political/diplomatic consequences for the United States unless the United States and China can cooperate on energy matters in such a way that it does not undermine U.S. and Chinese foreign policy goals in Angola.

UNITED STATES-CHINA-NIGERIA DIPLOMATIC/POLITICAL RELATIONS

U.S.-Nigerian diplomatic relations have improved considerably since Nigeria's restoration to democracy in 1999.[27] Developmentally, Nigeria has been plagued by poor governance, entrenched corruption, internal conflict, rampant health problems, and pervasive poverty.[28] Nigeria remains mired near the bottom of the United Nations Development Program (UNDP) human development index.[29] Regardless, Nigeria is America's largest exporter of petroleum in Africa, averaging 8 percent of total U.S. annual petroleum imports. Moreover, Nigeria is the fifth-largest exporter of oil to the United States globally. In addition, Nigeria is a member of OPEC, which as an organization has considerable influence over world oil prices. Therefore, Nigeria's stability and development is of the utmost importance to the United States because Nigeria is a major supplier of petroleum to the United States, and any political and economic instability in Nigeria would be detrimental to America's energy security.

Sino-Nigerian diplomatic relations are robust even though Nigeria is a very minor petroleum exporter to China. Diplomatic ties are largely driven by security cooperation and burgeoning bilateral trade, which are largely energy related, such as some hydrocarbon trading and Nigerian oil and infrastructure development. It should be noted that Nigeria is becoming one of China's largest trading partners in Africa when it comes to hydrocarbons.

In terms of Nigeria's oil and infrastructure development, China has concluded a number of oil agreements between November 2004 and December 2006 in which Chinese companies obtained oil exploration licenses and equity (partial ownership) in Nigeria oil fields in exchange for investments in Nigeria's infrastructure totaling $3.1 billion or more. Further, in April 2006, President Abuja offered four exploration licenses to Chinese oil firms for investments in Nigeria's infrastructure totaling $4 billion.[30] While China's approach to securing viable energy resources in Nigeria is viewed by many Western experts as contemptuous based on the fact that the Chinese oil companies have paid exorbitant prices in comparison for oil concessions and equity in comparison to their U.S. counterparts like ExxonMobil or Chevron, nevertheless China is building extensive relationships with the Nigerian government and elites, which essentially enables Nigeria to avoid any potential constraints placed on its conduct imposed by Western donor conditions.[31] Therefore, China's oil diplomacy potentially undermines Western and particularly U.S. efforts to maintain various good governance and anti-corruption initiatives in Nigeria through substantive international financial aid from the IMF and the World Bank. Further, China presents itself as an alternative political and economic model or as an exemplar of development without democracy to Nigeria vis-à-vis the West and the United States.

Although China's oil diplomacy has been generally favorable to Nigeria, it has been problematic. Chinese oil companies increasingly face armed reprisals from Nigerian militants of the Movement for Emancipation of Niger Delta (MEND) for the alleged Chinese expansion in the region, or Chinese "colonialism." Many militants feel that this region has been exploited for the benefit of Nigerian elites and foreign oil companies to the detriment of the Nigerian people. MEND militants have responded accordingly by attacking oil installations and kidnapping foreign oil workers, including Chinese nationals, in early 2006, leading to a 20 percent reduction in Nigeria's oil production.[32] Consequently, Nigeria turned to China for military supplies to protect the oil fields, claiming that the United States was tardy in its response to the destabilization of the security situation in the Delta.[33]

UNITED STATES-CHINA-SUDAN DIPLOMATIC/POLITICAL RELATIONS

U.S.-Sudanese diplomatic relations have been strained since June 1967, following the outbreak of the first Arab-Israeli War, and were further strained by the assassination of U.S. Ambassador Cleo A. Noel and Deputy Chief of Mission Curtis G. Moore in Khartoum by the Black September terrorist organization in March 1973, in which the Sudanese Government arrested and released the terrorists to the Egyptian Government; Sudan's support of Iraq in its invasion of Kuwait in 1990; providing sanctuary and assistance to Islamic terrorist groups and individuals such as Carlos the Jackal, Osama bin Laden, Abu Nidal, and other terrorist leaders who resided in Khartoum; and more recently by Sudan's continued complicity in unabated violence occurring in Darfur.[34] Consequently, the United States designated Sudan as a state sponsor of terrorism in 1993; suspended U.S. embassy operations in Khartoum in 1996; imposed comprehensive economic, trade, and financial sanctions against Sudan in October 1997, and launched cruise missile strikes against Khartoum in August 1998 in the wake of the East Africa embassy bombings.[35] Regardless, Sudan does not export petroleum to the United States. Sudan is a pariah state, in which China has made substantial investments in return for crude oil.

ECONOMIC IMPLICATIONS FOR THE UNITED STATES

China's interest in the Africa is primarily economic: energy trading, energy prospecting, and the trading of goods and services. In terms of energy trading, Africa is China's second largest crude oil and petroleum supplier after the Near East. Moreover, in terms of oil prospecting, China has ongoing upstream sector activities in Angola, Chad, Congo-Brazzaville, Cote d'Ivoire, Equatorial Guinea, Gabon, Mauritania, Nigeria, and Sudan and prospective upstream sector developments in Kenya, Madagascar, Mali, and Niger. Finally, in terms of goods and services, Africa is a major import market for Chinese raw and manufactured goods such as textiles, apparel, footwear, machinery and transportation equipment, ores and metals, agricultural raw materials, manufactured materials, and processed food and beverages.

China's principal strategic organ for developing and sustaining its economic relations (as well as its political relations) with Africa is the Forum on China-Africa Cooperation (FOCAC). Created in Beijing in 2000, the FOCAC was instrumental to not only advancing China's economic (and political) interests in Africa but also cementing its economic (and political) influence.

Through the FOCAC China was able to spawn further economic development and influence in Africa. At the 2006 FOCAC summit, China initiated the creation of Special Economic Zones (SEZs), which will provide liberalized investment environments focused on strategic industries to attract Chinese companies.[36] One such designated SEZ is Nigeria because Nigeria is a growing investment destination for Chinese oil companies like Sinopec, PetroChina (a CNPC subsidiary) and CNOOC. Nigeria is also America's largest supplier of petroleum in Africa and is the largest destination of U.S. and Western foreign investment in terms of oil exploration and development. Because Nigeria is heavily reliant on joint venture operations to fund oil exploration and development, and Chinese investors have been more willing to offer vitally needed infrastructure developments in exchange for drilling rights, Nigeria may choose Chinese oil companies over U.S. oil companies to drill for oil in the near future once drilling licenses come up for renewal.[37] Thus, Chinese excursions may result in resource nationalism. Chinese oil investment opportunities in Nigeria makes Nigeria is a potential battleground for U.S.-Sino economic competition.

China's oil energy trading and investment opportunities in Africa in the same countries in which the United States has similar interests may also drive up global oil prices. Because the United States and China are competing for crude oil and petroleum in the same oil-producing countries, and the world oil supply is finite but not too finite, the laws of supply and demand will most certainly dictate an increase in prices to meet the global energy demand, especially from the largest oil-consuming countries like the United States and China. Other factors have contributed to price fluctuations in the past, such as war and conflict and the global economy, but competition provides OPEC with the opportunity to manipulate global oil prices.

On a positive note, China's hydrocarbon trading and investment opportunities in Africa's high-risk countries where the United States and the West do not currently have oil investments could add new capacity to the world's tight energy market and therefore drive global oil prices down. In this respect,

China's direct commercial engagement with pariah states like Sudan, which has abundant crude oil reserves, actually may help to reduce global oil prices somewhat because China has no direct economic competition with either the United States or the West because U.S. and Western oil companies are statutorily precluded from commercially engaging Sudan. However, the Sino-Sudanese economic liaison has negative political ramifications for the United States.

In summary, China's oil energy and investments in Africa will not have negative economic consequences for the United States in the near future, notwithstanding substantial American aid, but as China becomes more actively involved in Africa's political and economic development through comprehensive measures, and the United States becomes more energy dependent on African petroleum resources, especially from Angola and Nigeria, it is very likely that a U.S.-Sino economic competition will emerge in the aftermath.

UNITED STATES-CHINA-ANGOLA ECONOMIC RELATIONS

U.S.-Angolan oil economic relations are good even though Angola exports approximately 3.5percent of America's total oil annually. U.S. oil companies are well represented and hold a dominant position in the Angolan oil production market vis-à-vis' the Chinese oil company Sinopec.

Sino-Angolan oil economic relations are robust as evidenced by the fact that Angola is China's largest petroleum energy trading partner in Africa. In this respect, Angola exports about 17percent of China's total annual petroleum, and China has oil equity stakes, offshore oil prospecting ventures, an oil refinery project under development, and oil-for-credit deals totaling $2 billion.

While China's petroleum trading and energy stakes in Angola pale in comparison to those of the United States and the rest of the West, China's approach to acquiring petroleum supplies through a "capital for oil" program may help Chinese oil companies like Sinopec to acquire the offshore drilling rights in the near future (which ExxonMobil subsidiary Esso and Chevron currently owns along with other oil majors like BP) once Sinopec develops significant technical knowledge in offshore drilling. Currently, Sinopec has joint offshore operations with the Angolan state-owned oil company Sonangol, which also supplies the technical expertise.

China's petroleum trading with Angola would certainly provide Angola with the opportunity to manipulate global oil prices for relative gains. In this respect, Angola is a member of OPEC, which has considerable influence over the production and price of oil. Angola, as a member of OPEC, could use

China's high demand for petroleum as an excuse to cut production and raise prices. Naturally, this action would affect the United States economically because the United States imports petroleum from Angola. Thus, China's petroleum trading and stakes in Angola would have negative economic consequences for the United States.

UNITED STATES-CHINA-NIGERIA ECONOMIC RELATIONS

U.S.-Nigerian oil economic relations are robust, as evidenced by the fact that Nigeria supplies the United States with 8percent of its total petroleum annually, making Nigeria the largest petroleum exporter in Africa to the United States, and that the United States is Nigeria's largest foreign investor. Most of U.S. FDI in Nigeria is concentrated in the petroleum/mining sectors, of which ExxonMobil and Chevron are the largest U.S. corporate players.[38] Therefore, Nigeria is vital to America's economic security as well as its energy security.

While Nigeria is currently a minor petroleum exporter to China, Sino-Nigerian oil economic relations are growing by virtue of the fact that Nigeria has been designated by China as a Special Economic Zone (SEZ), especially for Chinese oil companies. In this respect, China has ongoing investments in Nigeria's petroleum/mining sector, which includes equity stakes and oil exploration and development.

China's burgeoning investment in Nigeria's petroleum/mining sector—in which trade-related finance is tied to another major area of investment, infrastructure construction—is a matter of economic concern for the United States because China may potentially outdo U.S. competition by paying above-market prices to guarantee access to oil supplies. For example, China paid Nigeria $3.1 billion between 2004 and 2006 in infrastructure capital in return for offshore drilling rights. Further, China paid Nigeria $4 billion in April 2006 in return for four oil exploration licenses. Because Nigeria is heavily dependent on foreign direct investment to sustain its petroleum/mining-based economy, and China has been more willing to make sweeter infrastructure deals in exchange for drilling rights, Nigeria may choose to work with Chinese oil companies over American oil companies in the near future. This action would be potentially devastating to the U.S. oil industry given the high energy stakes that it has in Nigeria. Further, it is not an unlikely scenario given the corrupt nature of most, if not all, African regimes.

China's energy stakes in Nigeria could also contribute to a rise in global oil prices, which would have economic consequences for the United States. In this respect, Nigeria is a member of OPEC and a major oil supplier to the largest consumers of oil in the world: the United States, China and India. If Nigeria is supplying the world's three largest oil consumers as a member of OPEC, it could attempt to influence the price of oil globally, which would adversely affect the U.S. economy. Obviously Nigeria's estimated oil reserves pale in comparison to those of Saudi Arabia or the Gulf States. Nonetheless, Nigeria is a major player in global oil market and a potential bone of economic contention for the United States.

UNITED STATES-CHINA-SUDAN ECONOMIC RELATIONS

In general, U.S.-Sudanese economic relations, including oil relations, are cold even though Sudan has been the recipient of U.S. foreign aid and humanitarian assistance. Sudan has been economically sanctioned by the United States and the West because of its support for international terrorist networks. Further, Sudan is not a petroleum exporter to the United States.

Sino-Sudanese oil economic relations are strong as evidenced by the fact that Sudan exports approximately 31 percent of China's annual oil supply and that China has high hydrocarbon energy stakes in Sudan in terms of equity shares in oil fields, prospecting, and oil pipeline construction.

It is unlikely that Chinese petroleum energy trading and stakes in Sudan will have either short-term or long-term economic consequences for the United States because Sudan is not a petroleum exporter to the United States. Further, for the most part, past economic sanctions have limited Sudan's access to the global oil market, making Sudan's ability to influence global oil prices negligible. Therefore, China's hydrocarbon trading with high-risk states like Sudan may help to keep global oil prices down because China does not have to compete with the United States economically and does not have to abide by American-imposed economic sanctions, especially when it comes to petroleum trading.

MILITARY IMPLICATIONS FOR THE UNITED STATES

Although U.S. military engagement in Africa has been traditionally minimal, China's demand for African oil is unlikely to have negative military consequences for the United States or lead to a military competition for strategic energy resources in the short or medium term because Africa is not a critical part of China's national security. "Africa is a security issue only in terms of China's efforts to secure the energy, mineral and timber resources to fuel its economy."[39] Further, Chinese military activities (military assistance, arms transfers, and peacekeeping operations) are essentially a function of China's much greater use of soft power in advancing its greater diplomatic, economic, and security interests. However, if the oil-producing regions become politically unstable to the point where indigenous security forces cannot protect oil fields and foreign oil workers, then China may step in and provide military training, assistance, personnel, and equipment to protect its strategic oil energy interests. Angola, Cote d'Ivoire, Nigeria, and Sudan are the best examples in recent times where China has provided either military assistance or equipment to protect its oil energy interests. To Angola and Nigeria, China provided military training and weapons; to Cote d'Ivoire and Sudan China provided 13 and 613 peacekeepers respectfully under the UN umbrella. It should noted that China's most significant peacekeeping involvements in Africa are linked to countries where it has either a major interest in natural resources such as oil or there are special considerations.[40] Therefore, it appears that Chinese military activities in Africa are integral and subservient to China's larger national goals of securing vital energy resources rather than as an independent political actor.

China's demand for African oil will not have negative military consequences for the United States or lead to a military competition for strategic energy resources in the short or medium-term because of Africa's recent strategic importance to U.S. national security. In this respect, the United States (in comparison to China) has dramatically ramped up its military activities in Africa since 2002 because of the events of September 11, 2001. For example, under the Trans-Sahara Counter-Terrorism Initiative (TSCTI), the U.S. military has provided $500 million to increase border security and counter-terrorism capacity in Mali, Chad, Niger, and Mauritania. Further, the Combined Joint Task Force-Horn of Africa (CJTF-HOA) has been using Camp Lemonier in Djibouti as its main U.S. base for counterterrorist activities offshore in the region since 2002. In addition, the U.S. Navy has been patrolling the oil-rich Gulf of Guinea, a naval presence that was non-existent prior to 2005. Moreover, on October 1, 2008, the United States activated the U.S. Africa Command (AFRICOM), a Unified Combatant Command of the U.S. Department of Defense that is responsible for U.S. military operations and military relations with 53 African nations—an area of responsibility covering all of Africa except Egypt. Although AFRICOM is essentially military in nature, it seeks to better integrate U.S. civilian and military responses to security challenges in Africa, which includes energy security in those regions of Africa are that are more likely to experience terrorist attacks or political instability such as Angola and Nigeria, where the U.S. and China have its greatest energy interests. Although controversial, AFRICOM and other U.S. military assistance programs may actually help to improve energy security for all stakeholders, both foreign and domestic.

Although China historically has had good military relations with Africa, including being a major supplier of small arms and light weapons, it cannot compete with the United States militarily. China neither can project military power globally nor does it have operational experience. Further, China neither has military bases in Africa nor seeks traditional alliances as a matter of national policy.[41] Moreover, Chinese weaponry and military technology are generally inferior to those of the United States. Finally, China's military efforts are a corollary to Beijing's much greater use of soft power in advancing its strategic interests as opposed to the United States, which historically tends to rely on hard military power to advance its national interests. Therefore, it is unlikely that there will be a military competition for strategic energy resources between the United States and China in the short or medium term even though China's military presence and role in Africa are growing.[42]

In terms of military presence and influence in Africa, China has 14 of 107 worldwide military attaché offices, of which 2 host countries have proven oil reserves: Nigeria and Sudan. This is a modest statistic compared to U.S. military presence and influence in Africa. Further, between 2001 and 2006, China has conducted 30 major military exchanges and 110 bilateral security-related meetings and consultations; participated in 7 UN peacekeeping operations as of April 2008; and supplied arms to at least 16 African countries, of which 6 have real or potential oil reserves: Gabon, Kenya, Mali, Mauritania, Nigeria, and Sudan. Whether China's growing military presence and influence will result in a military competition for strategic energy resources in the long term remains to be seen. It all depends on whether China will resort to military force to accomplish its national political-economic goals.

UNITED STATES-CHINA-ANGOLA MILITARY RELATIONS

U.S.-Angolan military relations are developing even though Angola exports 4 percent of America's total petroleum annually. Angola has received U.S. military assistance in the form of $152,000 in International Military Education and Training (IMET) funds for English-language training to the Angolan armed forces and professional training for law enforcement personnel.[43]

Sino-Angolan military relations are growing even though Angola is China's primary source of oil in Africa. Angola is becoming a more important partner for military cooperation with China because Angola is China's primary source of oil in Africa.[44] Since 2000, China and Angola have conducted several high-level military visits, which have included President Hu Jintao and the

Chinese defense minister. Further, in 2007 China provided military technical assistance and supplied military equipment to the Angolan armed forces. In addition, Angola military personnel have received training in China.[45]

Although the United States and China have military relations with Angola, it does not appear that their oil energy interests in Angola will evolve into a military competition for Angola's oil energy resources in the short or medium term. There may be a high level of military cooperation with Angola, but it appears that this military cooperation has not materialized into a military rivalry between the United States and China.

UNITED STATES-CHINA-NIGERIA MILITARY RELATIONS

U.S.-Nigerian military relations are robust as are their energy relations. The United States maintains a permanent defense attaché in the capital city of Abuja and has conducted high-level military visits and meetings with Nigerian defense officials. Further, Nigeria has lent strong diplomatic support to U.S. counterterrorism efforts in the aftermath of the September 11, 2001. terrorist attacks and played a leading role in forging an anti-terrorism consensus among many states in Africa.[46]

Sino-Nigerian military relations are good even though Nigeria is a minor exporter of petroleum to China. Both countries have reciprocal defense attaché offices in their respective capitals. Further, both have conducted high-level military visits and meetings between 2000 and 2006. In addition, China has supplied Nigeria with small arms in 1997 and 1998 and fighter aircraft and air-to-air missiles in 2005.

Because of Nigeria's strategic value to the United States in terms of energy and national security and China's growing political, economic, and military presence, there is a potential for a military competition for Nigeria's oil resources between the United States and China. Nigeria is a key resource-rich African nation for both the United States and China in terms of oil.

UNITED STATES-CHINA-SUDAN MILITARY RELATIONS

Even though Sudan does not export petroleum to the United States, U.S.-Sudanese military relations are slowly being reestablished after strained relations in the 1990s. Sudan was a state sponsor of international terrorism but has provided concrete cooperation against international terrorism since the September 11, 2001, terrorism strikes on New York and Washington.[47]

Sino-Sudanese military relations are robust, as are their oil energy relations. Both countries have reciprocal defense attaché offices in their respective capitals. Further, both have conducted high-level military visits and meetings between 2000 and 2006. In addition, China supplied Sudan with small arms and fighter aircraft between 1997 and 2004. Finally, China was engaged in peacekeeping operations in Sudan under the UN flag.

It appears that a military competition for Sudan's oil energy resources will not evolve in the short or medium term between the United States and China because the United States has no oil energy stakes in Sudan.

Notes

1. Eisenman, Joshua, "China's Post-Cold War Strategy in Africa: Examining Beijing's Methods and Objectives," in Joshua Eisenman, Eric Heginbotham, and Derek Mitchell, eds., *China and the Developing World: Beijing's Strategy for the Twenty-First Century* (Armonk, New York: M.E. Sharpe), 2007, 35.

2. Eisenman, Joshua, "China's Post-Cold War Strategy in Africa: Examining Beijing's Methods and Objectives," 36.

3. Congressional Research Service, Library of Congress, *Comparing Global Influence: China's and U.S. Diplomacy, Foreign Aid, Trade, and Investment in the Developing World* (Washington, D.C.: U.S. Government Printing Office), RL 34620, August 15, 2008, 114.

4. Rotberg, Robert I., "China's Quest for Resources, Opportunities, and Influence in Africa" in Robert I. Rotberg, ed., *China Into Africa: Trade, Aid and Influence.* (Washington, D.C.: Brookings Institution Press), 2008, 1.

5. Eisenman, Joshua, "China's Post-Cold War Strategy in Africa: Examining Beijing's Methods and Objectives," 43.

6. Pham, J. Peter, "China's African Strategy and Its Implications for U.S. Interests," *American Foreign Policy Interests,* Issue 28, 2006, 247.

7. Eisenman, Joshua, "China's Post-Cold War Strategy in Africa: Examining Beijing's Methods and Objectives," 50.

8. Campos, Indira, Michael Chege, Pat Utomi, and Alex Vines, *U.S. and Chinese Engagement in Africa: Prospects for Improving U.S.-China Africa Cooperation,* (Washington, D.C.: CSIS Press), July 2008, 14.

9. See www.princeandpanda.com/media/6d4c7d4a7bb40f5bffff8019ffffa126.pdf.

10. Congressional Research Service, Library of Congress, *Comparing Global Influence: China's and U.S. Diplomacy, Foreign Aid, Trade, and Investment in the Developing World,* 118.

11. Pham, J. Peter, "China's African Strategy and Its Implications for U.S. Interests," *American Foreign Policy Interests,* 118 and Chietigj Bajpaee, "Sino-US Energy Competition in Africa," October 7, 2005, *www.globalpolicy.org/security/natres/oil/2005/1007competition.htm.*

12. Congressional Research Service, the Library of Congress, *China's Foreign Policy and "Soft Power" in South America, Asia, and Africa: A Study Prepared for*

the Committee on Foreign Relations United States Senate (Washington, D.C.: U.S. Government Printing Office), April 2008, 128 and Congressional Research Service, the Library of Congress, *Comparing Global Influence: China's and U.S. Diplomacy, Foreign Aid, Trade, and Investment in the Developing World*, 117–18.

13. Taylor, Ian, *China's New Role in Africa* (London: Lynne Rienner Publishers, Inc.), 2009, 20–21.

14. Wenran Jiang, "China's Emerging Strategic Partnerships in Africa" in Robert I. Rotberg, ed., *China Into Africa: Trade, Aid and Influence.* Washington, D.C.: Brookings Institution Press, 2008, 59.

15. Jiang, Wenran, "China's Emerging Strategic Partnerships in Africa" in Robert I. Rotberg, ed., *China Into Africa: Trade, Aid and Influence.* Washington, D.C.: Brookings Institution Press, 2008, 59.

16. Davies, Martyn J., "Special Economic Zones: China's Developmental Model Comes to Africa" in Robert I. Rotberg, ed., *China Into Africa: Trade, Aid and Influence.* Washington, D.C.: Brookings Institution Press, 2008, 142.

17. Davies, Martyn J., "Special Economic Zones: China's Developmental Model Comes to Africa", 142.

18. Congressional Research Service, Library of Congress, *Comparing Global Influence: China's and U.S. Diplomacy, Foreign Aid, Trade, and Investment in the Developing World*, 128.

19. Congressional Research Service, Library of Congress, *Comparing Global Influence: China's and U.S. Diplomacy, Foreign Aid, Trade, and Investment in the Developing World*, 128.

20. Davies, Martyn J., "Special Economic Zones: China's Developmental Model Comes to Africa", 142.

21. Congressional Research Service, Library of Congress, *Comparing Global Influence: China's and U.S. Diplomacy, Foreign Aid, Trade, and Investment in the Developing World*, 129.

22. Congressional Research Service, Library of Congress, *Comparing Global Influence: China's and U.S. Diplomacy, Foreign Aid, Trade, and Investment in the Developing World*, 129.

23. U.S. Department of State, Background Notes: Angola, *www.state.gov*, December 2008.

24. U.S. Department of State, Background Notes: Angola, *www.state.gov*, December 2008.

25. Lee, Henry and Dan Shalmon, "Searching for Oil: China's Oil Strategies in Africa" in Robert I. Rotberg, ed., *China Into Africa: Trade, Aid and Influence.* (Washington, D.C.: Brookings Institution Press), 2008, 118.

26. Lee, Henry and Dan Shalmon, "Searching for Oil: China's Oil Strategies in Africa", 123.

27. U.S. Department of State, Background Notes: Nigeria, *www.state.gov*, December 2008.

28. U.S. Department of State, Background Notes: Nigeria, *www.state.gov*, December 2008.

29. U.S. Department of State, Background Notes: Nigeria, *www.state.gov*, December 2008.

30. Taylor, Ian, *China's New Role in Africa* (London: Lynne Rienner Publishers, Inc.), 2009, 47–48.

31. Obiorah, Ndubisi, Darren Kew and Yushf Tanko, " 'Peaceful Rise' and Human Rights: China's Expanding Relations with Nigeria" in Robert I. Rotberg, ed., *China Into Africa: Trade, Aid and Influence.* (Washington, D.C.: Brookings Institution Press), 2008, 289.

32. Taylor, Ian, *China's New Role in Africa* (London: Lynne Rienner Publishers, Inc.), 2009, 49.

33. Taylor, Ian, *China's New Role in Africa* (London: Lynne Rienner Publishers, Inc.), 2009, 49.

34. U.S. Department of State, Background Notes: Sudan, *www.state.gov*, January 2009.

35. U.S. Department of State, Background Notes: Sudan, *www.state.gov*, January 2009.

36. Davies, Martyn J., "Special Economic Zones: China's Developmental Model Comes to Africa", 137.

37. Hurst, Cindy, *China's Oil Rush In Africa* (Washington, D.C.: Institute for the Analysis of Global Security), July 2006, 11.

38. U.S. Department of State, Background Notes: Nigeria, *www.state.gov*, December 2008.

39. Shinn, David H., "Military and Security Relations: China, Africa, and the Rest of the World" in Robert I. Rotberg, ed., *China Into Africa: Trade, Aid, and Influence.* (Washington, D.C.: Brookings Institution Press), 2008, 154.

40. Shinn, David H., "Military and Security Relations: China, Africa, and the Rest of the World," 177.

41. Shinn, David H., "Military and Security Relations: China, Africa, and the Rest of the World," 180.

42. Shinn, David H., "Military and Security Relations: China, Africa, and the Rest of the World," 183.

43. U.S. Department of State, Background Notes: Angola, *www.state.gov*, December 2008.

44. Shinn, David H., "Military and Security Relations: China, Africa, and the Rest of the World," 165.

45. Shinn, David H., "Military and Security Relations: China, Africa, and the Rest of the World," 165.

46. U.S. Department of State, Background Notes: Nigeria, *www.state.gov*, December 2008.

47. U.S. Department of State, Background Notes: Sudan, *www.state.gov*, January 2009.

Chapter 7

Western Europe

In terms of the states that produce crude oil for both the United States and China, Western Europe is comprised of Norway and the United Kingdom. Many observers do not see Western Europe as an area of strategic Sino-American competition for oil energy resources because the region has approximately 1 percent of the world's proven oil reserves. Further, the region tends align itself politically, economically, and militarily with the United States. In terms of U.S. crude oil imports, Western Europe produces approximately 3.1 percent. Comparatively, China imports less than 0.2 percent of its crude oil from Western Europe. Thus, Western Europe ranks last in the world as an oil producer to China.

The PRINCE model predicts that China will continue to import oil from Western Europe based on a probability of 90.9 percent or greater.[1]

POLITICAL, ECONOMIC, AND MILITARY IMPLICATIONS FOR THE UNITED STATES

Sino-Western European diplomatic relations have improved steadily since the end of the Cold War as evidenced by high-level political exchanges and regular meetings including political summits involving the national leaders of China and the European Union (EU). Further, Western Europe and China have engaged in economic, energy, environmental, scientific and technological, and educational and legal cooperation to strengthen bilateral ties. However, China and Western Europe have yet to forge a strategic partnership. Persistent concerns over China's authoritarian nature, legal system, and human rights record impede further development and deepening of bilateral diplomatic relations despite some considerable progress over the last 20 years, especially since June 1989 when China cracked down on political activists at Tiananmen Square, which resulted in an arms embargo on China and tensions in bilateral relations.

Western Europe's close ties with the United States also factor in the development of Sino-Western European diplomatic relations. Western Europe is aligned with the United States, having a long tradition of strategic cooperation that extends beyond the political realm and into the economic and military ones and shared commitments to a number of basic values that transcend the Sino-Western European relations despite political tensions stemming from America's past unilateralist approach to international relations, such as the invasion of Iraq and the treatment of unlawful combatants.[2] Therefore, China's energy relations with Western Europe, which mainly consists of importing crude oil from Norway and the United Kingdom, is unlikely to result in a Sino-American diplomatic competition for oil energy resources.

Trade and economic ties appear to be the foundation of the Sino-Western European relationship even though China has sought to foster "greater European political and strategic independence from the United States."[3] Western Europe (under the auspices of the European Union) has been China's largest trading partner, and the European Union has been China's second largest trading partner after the United States since 2004.[4]

Although trade and economic relations have deepened between China and Western Europe (under the auspices of the European Union), it has not been without friction.[5] Since late 2006, trade and economic problems have played a more prominent role in Sino-Western European economic relations across a wide range of issues: EU trade deficit, unfair trade practices, intellectual property rights violations, currency valuation, and the export of dangerous and tainted goods by China.[6] Further, Europe also refused to extend Market Economy Status (MES) to China because of unfair trade practices even though China joined the World Trade Organization in 2001.[7] Apart from persistent political issues in the relationship, such as China's governability and human rights record, these other issues have prevented China and Western Europe from implementing a strategic economic relationship.[8]

Western Europe's strategic relationship with the United States also factors in Sino-Western European economic relations. Despite past differences in foreign policy approaches, Western Europe and the United States have deeply integrated economies (not least in the defense sector) and shared commitments to a number of basic values.[9] China's energy relations with Western Europe, which mainly consists of importing crude oil from Norway and the United Kingdom, is unlikely to result in a Sino-American economic competition for oil energy resources. The level of oil energy trading between China and Western Europe, compared to the level of energy trading between the United States and Western Europe, also factors in the potential Sino-American economic competition for energy resources and will be addressed separately below.

As with political and economic relations, China's oil energy trading in Western Europe (which is marginal at best) is unlikely to lead to a military competition for strategic energy resources because of the strong military relationships that exist between the United States and the nations of Europe. The United States has had a far longer and deeper military relationship with Western Europe than China, especially since 1949, when the North Atlantic Treaty Organization (NATO) was formed for the collective defense of its members, including the United States, in the event of an attack by an external party. Participation as a member in NATO means participation in hard military activities such as coalition warfare, as opposed to soft military activities such as military attaché exchanges, educational exchanges, port visits, observing military exercises, or even UN peacekeeping. Note that historically China has never subscribed to formal military alliances, even with its erstwhile communist ally Russia. China has provided military aid and support to regimes that are inimical to U.S. interests, such as Burma, Iran, and Sudan. However, these actions have not translated into a formal military alliance. Overall, Western Europe is formally aligned to the United States militarily.

China's oil energy trading in Western Europe is unlikely to lead to a military competition for strategic energy resources because Western Europe is not vital to China's energy security compared to the Near East and Africa. In this respect, China imports less than 0.2 percent of its oil from Western Europe compared to Africa and the Near East, which are respectively 30 percent and 50 percent. Thus, Western Europe is strategically insignificant in terms of energy security compared to the other oil-producing regions of the world.

Notes

1. See www.princeandpanda.com/media/6d4c7d4a7bb40f5bffff8019ffffa126.pdf.
2. Small, Andrew, "The US Factor," in Stanley Crossick and Etienne Reuter, eds., *China-EU: A Common Future* (Singapore: World Scientific Publishing), 2007, 74.
3. Sutter, Robert G., *Chinese Foreign Relations: Power and Policy Since the Cold War* (Landam, Maryland: Rowman & Littlefield Publishers), 2008, 344–45.
4. Sutter, Robert G., *Chinese Foreign Relations: Power and Policy Since the Cold War*, 344–45.

5. Gill, Bates and Melissa Murphy, *China-Europe Relations: Implications and Policy Reponses for the United States* (Washington, D.C.: CSIS Press), May 2008, 10.

6. Gill, Bates and Melissa Murphy, *China-Europe Relations: Implications and Policy Reponses for the United States*, 10–11.

7. Shambaugh, David, Eberhard Sandschneider, and Zhou Hong, eds., *China-Europe Relations: Perceptions, Policies and Prospects* (New York, New York: Routledge, 2008), 136.

8. Cameron, Fraser and Yongnian Zheng, "Key Elements of a Strategic Partnership", in Stanley Crossick and Etienne Reuter, eds., *China-EU: A Common Future.* (Singapore: World Scientific Publishing), 2007, 6.

9. Small, Andrew, "The US Factor," 74.

Chapter 8

Western Hemisphere

In terms of the energy-producing states for the United States and China, the Western Hemisphere region is made up of Argentina, Bolivia, Brazil, Columbia, Cuba, Ecuador, Guatemala, Peru, Venezuela, and Canada. Many observers see this region as a potential area of Sino-American competition because both China and the United States have overlapping diplomatic, commercial, and security interests, and it is in America's backyard. In terms of U.S. hydrocarbon energy trading, crude oil and petroleum imports from the Western Hemisphere rank first in the world, with Canada and Mexico together being America's largest suppliers of fossil fuels.

Comparatively, China has imported less crude oil and petroleum from the Western Hemisphere than any other major region in the world other than the Asia-Pacific and Central Eurasia. China imported almost 7 percent of its total oil from the Western Hemisphere in 2007. The PRINCE model predicts that China will continue to explore for oil in and import oil from the Western Hemisphere based on a probability of 90.3 percent or greater.[1]

POLITICAL IMPLICATIONS FOR THE UNITED STATES

China's growing interest in the Western Hemisphere, specifically Latin America and the Caribbean, is a fairly new phenomenon that has developed in recent years.[2] Using soft power and dollar diplomacy as its primary diplomatic tools, China's strategic agenda in the region is primarily driven by economic interests such as trade with the strategically economic states of Argentina, Brazil, Chile, and Mexico, although China has also noticeably intensified its diplomatic engagement with the region in recent years because of political competition with Taiwan.[3] In this respect, China desires to politically isolate Taiwan and obtain diplomatic recognition from those countries that currently recognize Taiwan in the Caribbean, Central America, and South America: El Salvador, Guatemala, Honduras, Nicaragua, Panama, Belize, Dominican Republic, Haiti, St. Kitts and Nevis, St. Lucia, Vincent and the Grenadines, and Paraguay. It should be noted that of the 12 countries in the region that formally recognize Taiwan, only one—Guatemala—sporadically exports oil to China.

China also seeks to strengthen its diplomatic relations with the resource-rich nations of the Western Hemisphere to help feed its resource needs and expanding economy.[4] Countries such as Canada, Colombia, and Venezuela increasingly play important roles in China's energy security as China seeks to find reliable and secure petroleum resources in the region. Mexico could potentially play an important role in China's energy security but cannot at present because the Mexican constitution prohibits Chinese oil companies from acquiring the rights to explore for and produce oil, and Petroleos Mexicanos (the Mexican state-owned oil monopoly) would not be able to increase its crude oil production to meet China's current energy needs.[5]

Cuba will play an important role to China's energy security in the near future because of a recent discovery of 100 million barrels of oil in the Gulf of Mexico next to Cuba. Because U.S. oil companies are prohibited from drilling in waters 100 miles from Florida, Chinese oil companies can explore for and produce crude oil in the Gulf of Mexico, which is in America's backyard. In this respect, Cuba's state oil company Cubapetroleo (CUPET), signed a production sharing agreement with Sinopec after Fidel Castro's announcement of the oil discovery. Such a commercial venture, in conjunction with China's established trade/intelligence relationship with Cuba, will increase China's presence in Cuba, which is approximately 90 miles from the United States.[6]

Although China has been careful not to antagonize the United States—given that the United States is sensitive to Chinese involvement in its neighborhood and in spite of the fact that U.S. linkages with the region are far greater, largely because of geographic proximity and extensive historical and cultural ties—China poses a potential threat to U.S. influence and interests in the region.[7] According to the Congressional Research Service, "some (analysts) maintain that by presenting an alternative political and economic model—rapid-state-sponsored economic growth and modernization alongside political authoritarianism—the PRC undermines the U.S. agenda to advance political reform, human rights and free trade in the region."[8] Second, according to some analysts, China's regional presence ultimately could have strategic implications for the United States in the event of a possible military conflict with China.[9] China could use its human and commercial infrastructure in the region to disrupt and distract the United States in the hemisphere. [10] China has already has an intelligence relationship with Cuba in which China has been operating two intelligence signal facilities that include cyber warfare units that monitor computer data traffic from the United States and a large complex equipped with 10 satellite communications antennas that intercepts U.S. telephone communications.[11]

China's hydrocarbon energy trading with the Western Hemisphere may lead to a Sino-American diplomatic competition for oil energy resources, especially in those countries that appear to have some ideological affinity with China: Argentina, Brazil, and Venezuela.[12] In terms of U.S. energy security, Venezuela is the most important country in the Western Hemisphere after Canada and Mexico because Venezuela exports approximately 70 percent of its oil to the United States and therefore is America's largest supplier of oil in South America. China's hydrocarbon energy trading with Venezuela may result in a strategic energy partnership, which could lead to a change in Venezuela's energy policy that increasingly favors gradual diversification away from the U.S. market or even a major supply disruption if U.S.-Venezuelan diplomatic relations sour. Overall, such scenarios warrant further examination country by country of how China's crude oil and petroleum energy trading and activities in the Western Hemisphere may have diplomatic implications for the United States.

UNITED STATES-CHINA-VENEZUELA DIPLOMATIC/ POLITICAL RELATIONS

Although U.S.-Venezuelan diplomatic relations have been tense in recent years, oil energy trading is strong and deep. Approximately 70 percent of Venezuela's oil exports are sent to the United States annually, making the United States Venezuela's most important oil energy trading partner.[13] According to Sidney Weintraub there are two reasons: the ability of U.S. refineries to handle heavy crude from Venezuela and the relatively low shipping cost, compared to the shipments to China.[14] Moreover, Venezuela is America's third largest supplier of petroleum in the Western Hemisphere after Canada and Mexico. In this respect, Venezuela produces approximately 10 percent of America's petroleum annually compared to Canada's 18 percent or

Mexico's 11 percent. Finally, Venezuela's economic prospects and President Chavez's political longevity are highly dependent not only on oil prices but also the export of petroleum. Therefore, despite President's Chavez incendiary ideological rhetoric and threats to cut off America's oil supply, the United States remains the single most important destination of Venezuela's crude oil. President Chavez cannot afford to instigate an oil embargo toward the United States without risking his political future.

Sino-Venezuelan diplomatic relations have developed smoothly since 1974, when formal diplomatic ties were established, as evidenced by the numerous high-level official visits. President Chavez made four official state visits in the last ten years (1999, 2001, 2004 and 2006), which symbolizes Venezuela's desire to reach out to expand diplomatic relations beyond the United States.

Sino-Venezuelan oil energy relations are also strong, as evidenced by the fact the Venezuela produces 4–5 percent of China's oil annually and that China has prospective oil upstream (exploration and production) ventures in Venezuela. Recently, China and Venezuela have agreed to build a 400,000 barrel-per-day joint venture refinery in South China's Guangdong Province, which will advance President Chavez's goal of shipping to China 1 million barrels per day of oil by around 2011, or 13 percent of current Chinese oil demand. Currently, Venezuela ships nearly 300,000 barrels per day of oil and fuel to China, of which around 80,000 barrels are crude oil.

China's access to Venezuela's oil energy market sparks U.S. nervousness because of the potentially serious diplomatic implications of a China-Venezuela energy relationship. A Sino-Venezuelan energy relationship could mean gradual market diversification from the United States and toward China or even a major supply disruption, which would threaten U.S. energy security. Nevertheless, it is apparent that there is a triangular aspect to Sino-Latin American-U.S. energy relations in the case of Venezuela, which could result in a Sino-American diplomatic competition for Venezuela's oil energy resources in the near term. So if China desires to deepen its oil energy ties with Venezuela without diplomatically/politically provoking the United States, it may be in China's interest to pressure Venezuela to tone down its anti-U.S. rhetoric. Otherwise, it would be more costly for the Chinese government.

UNITED STATES-CHINA-CANADA DIPLOMATIC/POLITICAL RELATIONS

U.S. diplomatic relations with Canada are the closest and most extensive in the world, according to the U.S. State Department.[15] The United States and Canada cooperate on multiple levels (from federal to local government) and many forums, including national security, law enforcement, counterterrorism, counter-narcotics, the environment, and energy.

In terms of oil energy trading, Canada is America's largest supplier of oil at 18 percent annually. making Canada the most important supplier of oil to the United States and also the most secure supplier in the hemisphere—and perhaps in the world—because of the friendship and substantial cooperation between the two countries. [16] More than 99 percent of Canada's crude oil exports are sent to the United States.[17]

Sino-Canadian diplomatic relations have generally been very good, even though Canada produces less than 1 percent of China's crude oil annually. However, since Prime Minister Stephen Harper and his Conservative Party won the election in January 2006, relations have cooled.[18] Canada offended China by criticizing its human rights record, awarding honorary Canadian citizenship to the Dalai Lama, accusing China of commercial espionage, delaying meetings between foreign ministers, and making favorable comments toward Taiwan.[19]

China's oil energy stakes in Canada will not result in a Sino-American diplomatic competition for oil energy resources because of the U.S.-Canadian strategic relationship. In this respect, the U.S. and Canada are natural allies on many issues and multiple levels. Further, Canada is America's largest (and most reliable) supplier of crude oil in the world as evidenced by the fact the more than 99 percent of Canadian's crude oil exports are sent to the United States. However, should the relationship change between the United States and Canada where China would be able to cut into America's oil supply from Canada, then there would be Sino-American diplomatic competition for oil energy resources.

ECONOMIC IMPLICATIONS FOR THE UNITED STATES

Although the People's Republic of China seeks diplomatic recognition as the legitimate representative of China in international venues, especially with those countries in the Western Hemisphere that formally recognize Taiwan and desires to strengthen its political alliances in international and trade forums, China's primary interest in the Western Hemisphere—especially in Latin America—appears to be economically motivated, with China eager for access to such commodities as iron and other ores, soybeans, copper, iron and steel, integrated circuits and other electrical machinery, and oil to meet the demands of its booming economy and access to regional markets for its manufactured goods.[20] The Western Hemisphere—especially Latin America—is rich in energy, agricultural and mineral resources, and potentially serves as a major supplier to China in this respect, which may have strategic economic implications for the United States in terms of energy security. In this respect, the majority of America's oil and petroleum supplies excluding Mexico—approximately 33 percent—originate from the Western Hemisphere.

For the Western Hemisphere, especially Latin America, China is an ideal source for infrastructure investment because many Latin American countries suffer from poverty and indebtedness, to which the United States has devoted little attention. In this respect, "the lion's share of funds flowing from the United States to Latin America is targeted toward fighting drug production and trafficking."[21] Further, it appears that China's investment in Latin America may include future infrastructure improvement based on President Hu's 2004 controversial statement about a $100 billion 10-year investment for Latin America.[22] Thus, many Latin American countries are increasingly viewing China not only as a key trading partner but also as a potential source of much-needed infrastructure investment.[23] Moreover, in terms of overall trade, Latin America economically benefits from trade diversification, foreign direct investment, low-cost imports, and growth in sectors that are complementary to China's trade with the region.[24]

If China is able to further develop its bilateral trade ties in the Western Hemisphere, especially in Latin America, then there would be significant economic consequences for the United States. However, the volume of trade between China and the Western Hemisphere is relatively small compared to that of the United States. In addition, the United States is the largest trading partner to most of the countries in the Western Hemisphere even though Sino-Western Hemispheric trade has increased substantially in recent years. Moreover, China's investments in the region are relatively small compared with its investments elsewhere in the world.[25] In this respect, China is not a major foreign investor in the region. The European Union (EU) is the largest source of FDI in Latin America, with $620 billion invested in the region in 2006, followed by the United States with $350 billion.[26]

Additionally, China is geographically distant to the Western Hemisphere compared to the United States. Because Canada and Latin America are located in the same hemisphere as the world's superpower, they have been traditionally secure regions for the United States. Finally, the Western Hemisphere, especially Latin America, "is not as high a priority for China when compared to other more developed countries or those in its regional sphere."[27] Nevertheless, China is a potential economic rival to the United States in the Western Hemisphere, especially in Latin America.

China's oil energy trading and investment opportunities in the Western Hemisphere may have strategic economic implications for the United States.

First, it may drive up global oil prices. Second, it could result in an economic competition for oil energy resources, especially in those countries where the United States and China receive a substantial portion of their hydrocarbons: Brazil, Ecuador, and Venezuela. Canada and Mexico are excluded because the former exports the majority of its oil—more than 99 percent—to the United States, and the latter constitutionally prohibits China from oil investing and prospecting. Economic competition could be manifested in the form of trading blocs that could exclusively favor one country over another. China has already joined Brazil, India, and Russia in the formation of a group of four emerging economic powers known as BRIC. Further, China is developing a strategic energy partnership with Venezuela, one of the United States' major suppliers of petroleum in the Western Hemisphere. Third, it may spur a condition known as resource nationalism, whereby countries nationalize their energy resources by commandeering any or all existing oil equities, ceasing all domestic prospecting operations by independent oil companies and hastily expelling independent oil companies. This condition is highly possible especially in the politically instable countries of Colombia, Peru, and Venezuela. Although the Western Hemisphere is a minor oil producer to China at almost 7 percent annually, China's oil energy trading and investment in the region will have economic consequences for the United States. It behooves the United States to develop and implement a new economic strategy to counter any China's petroleum energy ventures in the Western Hemisphere.

UNITED STATES-CHINA-VENEZUELA ECONOMIC RELATIONS

U.S.-Venezuelan economic relations are robust despite political tensions. The United States is Venezuela's largest trading partner. Venezuelan exports to the United States were $40 billion (accounting for at least 60 percent of total Venezuelan exports) in 2007, and United States exports to Venezuela were $10 billion (or 22 percent of total Venezuelan imports) in that year.[28]

In terms of energy trading, Venezuela is the United States' largest supplier of oil in South America and fourth largest supplier after Canada, Mexico, and Saudi Arabia. Almost 70 percent of Venezuela's oil is exported to the United States.

Sino-Venezuelan economic relations are generally good but small compared to the United States. China exports electromechanical equipment, shoes, textiles, and garments and imports iron ore and finished oil from Venezuela.

In terms of energy trading, China imports approximately 5 percent of its oil from Venezuela and has burgeoning investments in Venezuela's upstream sector, including equity in at least two exploration blocks.

China's oil energy stakes in Venezuela could potentially result in a Sino-American economic competition for oil energy resources in the form of an energy trading bloc, which would favor China or gradual diversification away from United States given Venezuela's persistent anti-U.S. rhetoric and developing trade relations with China. However, Venezuela would have to weigh any adverse economic actions against its trade relationship, especially since the United States is Venezuela's top trading partner. In this respect, the Venezuelan economy is heavily dependent on oil revenues, and almost 70percent of its oil is exported to the United States. Further, only U.S. refineries can accommodate Venezuela's heavy crude oil, not China's. China and Venezuela have a formal agreement to build an oil refinery in Guangzhou to refine Venezuela's heavy oil, but the project will take some time before it matures. Nevertheless, although Venezuela has much to lose economically if it wages an oil energy trading war against the United States, the notion of growing Sino-Venezuelan energy ties probably is the most contentious issue in Sino-U.S. relations in the Western Hemisphere.

UNITED STATES-CHINA-CANADA ECONOMIC RELATIONS

U.S.-Canadian economic relations are robust. as evidenced by the fact that both nations share the world's largest and most comprehensive trading relationship, which supports millions of jobs in each country.[29] Total trade between the two countries exceeded $560 billion in 2007, of which the United States accounted for more than two-thirds of Canada's overall trade.[30]

In terms of investment, the United States is Canada's largest foreign investor.[31] In terms of energy trading, the United States and Canada enjoy the largest energy trade relationship in the world, according to the U.S. State Department.[32] Canada is the single largest foreign supplier of oil in the world, producing 18 percent annually—approximately 2.9 million barrels daily. More than 99 percent of Canada's oil is exported to the United States.

Sino-Canadian economic relations are good even though Canada produces less than 1 percent of China's oil annually. However, compared to the United States, Canada's trade with China is extremely small. In 2007, China accounted for 6 percent of Canada's total trade compared to that of the United States, which was 66 percent. Further, China's bilateral foreign direct investment in Canada is low compared to that of the United States. For example, in 2007, while almost 49 percent of Canada's FDI went to the United States, only 0.3 percent went to China.[33] Moreover, 58 percent of Canada's overall FDI was from the United States, while China's share was 0.1 percent of the total.[34]

China's oil energy investments in Canada will not result in a Sino-American economic competition for oil energy resources because the U.S.-Canadian economic ties are strategic in nature. Sino-Canadian trade relations are extremely small compared to U.S.-Canadian trade relations. Further, Canada is the United States' single largest supplier of oil in the world.

MILITARY IMPLICATIONS FOR THE UNITED STATES

China's oil energy trading and investments in the Western Hemisphere are unlikely to shift the balance of military power and influence from the United States to China or lead to a military competition for strategic energy resources because of the strong military relationships that exist between the United States and the countries of Western Hemisphere with the exception of Cuba. The United States has had a far longer and deeper defense relationship with the Western Hemisphere and remains its main arms provider.[35] Further, the Western Hemisphere is not vital to China's energy security compared to the Near East and Africa. In this respect, China imports only less than 7 percent of its oil from the region while it imports respectively approximately 30 percent from the Western Hemisphere and almost 50 percent from Africa and the Near East.

Although Chinese military presence in the Western Hemisphere is marginal compared to that of the United States, it is rapidly emerging as a military player in the region.[36] According to CSIS, "China is using military diplomacy, educational exchanges and medical and technical assistance in a measured way to support its broader political and economic interests in Latin America."[37] Obviously, greater access to natural resources such as oil fits into the broader political and economic scheme of China's foreign relations in the Western Hemisphere and will keep the Chinese military engaged in the region as part of a comprehensive effort to build mutually beneficial economic development, trade, military links, and political relations rooted in principles of South-South Cooperation similar to China's efforts in Africa. Thus, the Chinese military or the People's Liberation Army (PLA) in reality operates increasingly as an instrument of diplomatic statecraft of the People's Republic of China.[38] The PLA's role "constitutes just one part of an expanding presence that China manifests as a 'major power' on the global stage."[39] Further, "the PLA's involvement in Latin America illustrates China's pursuit of a multi-faceted strategy to expand its global presence."[40]

In terms of military presence in the Western Hemisphere, China has nine defense attaché offices, for which all host countries have proven oil reserves: Argentina, Belize, Brazil, Chile, Colombia, Cuba, Ecuador, Mexico, Peru, and Uruguay. Of these, only four do not produce oil for China: Belize, Chile, Mexico, and Uruguay. In addition to these official links, the Chinese military is engaged in Bolivia, Paraguay, and Venezuela through high-level military visits and arms sales. The fact that the Chinese military has a growing presence in the Western Hemisphere is disconcerting for the United States because it reveals that traditional U.S. military links with sister armed services in the region may have deteriorated somewhat since September 11, 2001, making room for Chinese military involvement to become more visible and effective.[41] In this respect, some critics argue that U.S. military relations with Latin America have been largely driven by the global war on terror in recent years rather than by traditional military-to-military links. Thus, U.S. military ties with the region may be waning, which presents an opportunity for China to fill a military vacuum.

UNITED STATES-CHINA-VENEZUELA MILITARY RELATIONS

U.S.-Venezuelan military relations are cold even though Venezuela continues to be a major producer of oil to the United States. Venezuela neither participates in any American-sponsored multinational exercises nor has made any military weapons or equipment purchases from the United States since 2003. Further, Venezuela stopped sending personnel to the Western Hemispheric Institute for Security Cooperation in 2004.

Hugo Chavez's Venezuela is considered to be a threat to democracy in Latin America because of its military arms purchases (largely from Russia), its relations with Cuba and Iran, and its efforts to export its brand of populism to other Latin American countries.[42] In terms of oil energy, there is concern that Venezuela is looking to develop China as a replacement market.[43]

Sino-Venezuelan military relations are very good. Bilateral military ties include military and defense-related educational exchanges, frequent official visits by senior defense leaders, cultural events such as visits to Venezuela by the PLA band and acrobatic troupes, intelligence cooperation, and Venezuelan purchases of Chinese arms and equipment such as air radars, 24 K-8 basic trainers and ground attack aircraft, 10 Y-12 twin-turbo prop, short take-off and landing, general purpose transport aircraft, military bridges, pontoons, earth movers, and field kitchens in 2005.[44]

Although Venezuelan President Chavez has been highly critical of the United States, China's energy stakes in Venezuela are unlikely to result in a Sino-American military competition for oil energy resources because the United States is Venezuela's largest petroleum customer in terms of physical volume and sales. Further, only the United States can refine Venezuela's heavy crude oil. In addition, China does not wish to damage its overall relations with the United States by appearing to interfere in America's traditional sphere of political influence—the Western Hemisphere. In this respect, China is keenly aware of President Chavez's unpredictable nature and his predisposition to stir things up politically. However, China's growing military ties with Venezuela in conjunction with its pursuit of energy resources gives pause for concern, especially since it is in America's backyard, and Venezuela is the America's supplier of oil in South America.

UNITED STATES-CHINA-CANADA MILITARY RELATIONS

U.S.-Canadian military relations are strategic in nature. Both the United States and Canada share NATO mutual security commitments and cooperate on continental defense within the framework of the North American Aerospace Defense Command (NORAD).[45] Further, both countries participate in multinational exercises sponsored by the U.S. Southern Command that are related to security, counterterrorism, and peacekeeping. In addition, bilateral cooperation includes educational exchanges for commissioned officers at service schools (war colleges) and the Naval Postgraduate School in Monterey, California. Finally, the United States is a major supplier of military weapons and equipment, including attack aircraft, transport aircraft, helicopters and radars, according to SIPRI.

Sino-Canadian military relations are limited compared to those of the United States. Apart from permanent defense attaché representation, China and Canada do not participate in multinational exercises or educational exchanges. Further, neither supplies weapons or equipment to the other.

China's energy trading with Canada is unlikely to result in a Sino-American military competition for oil energy resources because the strength and durability of the U.S.-Canadian relationship on all levels. Further, Canada produces less than 1 percent of China's oil compared to the United States. In addition, 99 percent of Canada's oil and petroleum is exported to the United States, making the United States Canada's largest energy partner.

Notes

1. See www.princeandpanda.com/media/6d4c7d4a7bb40f5bffff8019ffffa126.pdf.

2. Congressional Research Service, Library of Congress, *China's Foreign Policy and "Soft Power" in South America, Asia, and Africa: A Study Prepared for the Committee on Foreign Relations United States Senate* (Washington, D.C.: U.S. Government Printing Office), April 2008, 16.

3. Roett, Riordan and Guadalupe Paz, "Introduction: Assessing the Implications of China's Growing Presence in the Western Hemisphere," in Riordan Roett and Guadalupe Paz, eds., *China's Expansion into the Western Hemisphere: Implications for Latin America and the United States.* (Washington, D.C.: Brookings Institution Press), 2008, 16.

4. Congressional Research Service, the Library of Congress, *Comparing Global Influence: China's and U.S. Diplomacy, Foreign Aid, Trade, and Investment in the Developing World* (Washington, D.C.: U.S. Government Printing Office), RL 34620, August 15, 2008, 158.

5. Teng, Chung-Chian, "Hegemony or Partnership: China's Strategy and Diplomacy Toward Latin America," in Joshua Eisenman, Eric Heginbotham, and Derek Mitchell, eds., *China and the Developing World: Beijing's Strategy for the Twenty-First Century.* (Armonk, New York: M.E. Sharpe), 2007, 97.

6. U.S. Congress, Subcommittee on the Western Hemisphere, *China's Influence on the Western Hemisphere*, Hearing on April 6, 2005, 109th Congress, Session 1, 2005, 55.

7. Congressional Research Service, Library of Congress, *Comparing Global Influence: China's and U.S. Diplomacy, Foreign Aid, Trade, and Investment in the Developing World*, 156.

8. Congressional Research Service, Library of Congress, *Comparing Global Influence: China's and U.S. Diplomacy, Foreign Aid, Trade, and Investment in the Developing World*, 156.

9. Congressional Research Service, Library of Congress, *Comparing Global Influence: China's and U.S. Diplomacy, Foreign Aid, Trade, and Investment in the Developing World*, 156.

10. Congressional Research Service, Library of Congress, *Comparing Global Influence: China's and U.S. Diplomacy, Foreign Aid, Trade, and Investment in the Developing World*, 156.

11. U.S. Congress, Subcommittee on the Western Hemisphere, *China's Influence on the Western Hemisphere*, Hearing on April 6, 2005, 109th Congress, Session 1, 2005, 46.

12. Teng, Chung-Chian, "Hegemony or Partnership: China's Strategy and Diplomacy Toward Latin America," 93.

13. Weintraub, Sidney, *The Energy Situation in the Western Hemisphere* (Washington, D.C.: CSIS Press), July 2007.

14. Weintraub, Sidney, *The Energy Situation in the Western Hemisphere* (Washington, D.C.: CSIS Press), July 2007.

15. U.S. Department of State, Background Notes: Canada, *www.state.gov*, November 2008.

16. Weintraub, Sidney, *The Energy Situation in the Western Hemisphere* (Washington, D.C.: CSIS Press), July 2007.

17. Weintraub, Sidney, *The Energy Situation in the Western Hemisphere* (Washington, D.C.: CSIS Press), July 2007.

18. www.cbc.ca/news/background/china/china-canada-relations.html.

19. www.cbc.ca/news/background/china/china-canada-relations.html.

20. Bliss, Catherine E., "China's Projection of Soft Power in the Americas", in Carola McGiffert, ed., *Chinese Soft Power and Its Implications for the United States* (Washington, D.C.: CSIS Press), March 2009, 45 and Congressional Research Service, the Library of Congress, *China's Foreign Policy and "Soft Power" in South America, Asia, and Africa: A Study Prepared for the Committee on Foreign Relations United States Senate*, 16.

21. Xiang, Lanxin, "An Alternative Chinese View," in Riordan Roett and Guadalupe Paz, eds., *China's Expansion into the Western Hemisphere: Implications for Latin America and the United States* (Washington, D.C.: Brookings Institution Press), 2008, 55.

22. Xiang, Lanxin, "An Alternative Chinese View," 55.

23. Xiang, Lanxin, "An Alternative Chinese View," 55.

24. Roett, Riordan and Guadalupe Paz, "Introduction: Assessing the Implications of China's Growing Presence in the Western Hemisphere," 16.

25. Bliss, Catherine E., "China's Projection of Soft Power in the Americas," 50.

26. Lum, Thomas, Hannah Fischer, Julissa Gomez-Granger, and Anne Leland, *China's Foreign Aid Activities in Africa, Latin America, and Southeast Asia* (Congressional Research Service, Library of Congress, R40361, February 25, 2009), 13.

27. Tokatlian, Juan Gabriel, "A View from Latin America," in Riordan Roett and Guadalupe Paz, eds., *China's Expansion Into The Western Hemisphere: Implications for Latin America and the United States* (Washington, D.C.: Brookings Institution Press), 2008, 61.

28. U.S. Department of State, Background Notes: Venezuela, *www.state.gov*, January 2009.

29. U.S. Department of State, Background Notes: Canada, *www.state.gov*, November 2008.

30. U.S. Department of State, Background Notes: Canada, *www.state.gov*, November 2008.

31. U.S. Department of State, Background Notes: Canada, *www.state.gov*, November 2008.

32. U.S. Department of State, Background Notes: Canada, *www.state.gov*, November 2008.

33. Tiagi, Raaj and Lu Zhou, *Canada's Economic Relations with China* (Canada: Fraser Institute), February 2009, 1.

34. Tiagi, Raaj and Lu Zhou, *Canada's Economic Relations with China*, 1.

35. Horta, Loro, "In Uncle Sam's Backyard: China's Military Influence in Latin America," *Military Review*, September-October 2008, 55.

36. Horta, Loro, "In Uncle Sam's Backyard: China's Military Influence in Latin America," *Military Review*, September-October 2008, 51.

37. Flanagan, Stephen J., "China's Military Diplomacy and Engagement in Latin America," in Carola McGiffert, ed., *Chinese Soft Power and Its Implications for the United States* (Washington, D.C.: CSIS Press), March 2009, 56.

38. Watson, Cynthia, "The PLA in Latin America," *China Brief*, (Jamestown Foundation), Volume VII, Issue 20, October 31, 2007, 1.

39. Watson, Cynthia, "The PLA in Latin America," *China Brief*, (Jamestown Foundation), Volume VII, Issue 20, October 31, 2007, 1.

40. Watson, Cynthia, "The PLA in Latin America," *China Brief*, (Jamestown Foundation), Volume VII, Issue 20, October 31, 2007, 1.

41. Watson, Cynthia, "The PLA in Latin America," *China Brief*, (Jamestown Foundation), Volume VII, Issue 20, October 31, 2007, 5.

42. Sullivan, Mark P., *Venezuela: Political Conditions and U.S. Policy* (Congressional Research Service, Library of Congress), RL32488, August 1, 2008, 33.

43. Sullivan, Mark P., *Venezuela: Political Conditions and U.S. Policy*, 45.

44. Horta, Loro, "In Uncle Sam's Backyard: China's Military Influence in Latin America," *Military Review*, September-October 2008, 55.

45. U.S. Department of State, Background Notes: Canada, www.state.gov, November 2008.

Chapter 9

Summary

Comprehensive regional analysis indicates that Sub-Saharan Africa and the Near East are the two most unstable regions of the world. This presents a vexing problem for China because 77 percent of its oil is imported from these regions combined, of which 100 percent is transported by sea along the vital sea lines of communication (SLOCs), the commercial shipping routes that the U.S. Navy and allied navies patrol constantly. Even though there are few conditions that are likely to bring a military confrontation between China and United States over oil energy resources, it is not inconceivable that the United States may need to come to China's assistance in the near future to keep open some of the sea lines of communication (SLOCs), such as the Bab-el-Mandeb on the Red Sea or the Straits of Hormuz in the Persian Gulf, in light of the fact that the bulk of China's imported oil passes through these vital chokepoints. What must also be taken into consideration is that Japan, the United States' foremost ally in Asia, imports a considerable amount of its oil from the Near East—approximately 78 percent, according to the U.S. Energy Information Administration. That fact may generate additional pressure on the United States to intervene in any conflict affecting the security of the chokepoints.

Through the adroit use of soft power, China's strategic engagement in Sub-Saharan Africa and the Near East for the pursuit of scarce energy resources not only undermines the United States' efforts to obtain reliable crude oil and petroleum sources but also its ability to promote democratization, good governance, and transparency reforms. Cases in point are Angola and Nigeria, where China undermined Western and U.S. efforts to maintain various good governance and anti-corruption initiatives through substantive financial aid. Although generally favorable in Nigeria, China's soft power has been problematic because local militants view Chinese presence as expansionism and have targeted Chinese oil companies and oil workers as a consequence. In Iran and Sudan, China's adroit petroleum relations have complicated U.S. efforts to enforce economic sanctions for human rights and nuclear non-proliferation violations.

China's hard power and geopolitical influence is growing steadily in tandem with its pursuit and development of scarce energy resources. Chinese warships are currently patrolling the Bab-el-Mandeb in support of international anti-piracy operations. China is also Iran's largest conventional weapons supplier and plays a major role in Iran's military modernization and nuclear development program. China has quietly engaged in a strategy of military and economic cooperation with Pakistan, Myanmar, Bangladesh, Sri Lanka, and the Maldives in the Indian Ocean region in order to build and/or have access to ports and airbases to support not only a future tanker fleet but also a blue-water navy capable of projecting power not only in the Indian Ocean region but in the Persian Gulf. Beyond the Indian Ocean region China has been offered the use of a former U.S. airbase in Ecuador.

Overall, China's strategic engagement and rising geopolitical influence in tandem with its adroit pursuit and development of scarce energy resources causes it to be a strategic energy resource competitor politically, economically, and militarily, to the United States in those countries where both nations obtain their petroleum products. China will alter the status quo within the next 50 to 100 years and surpass the United States as a global power unless it can find a way to reduce its energy dependence on hydrocarbon-based fuels.

Part II

The Surplus Scenarios

Chapter 10

Surplus Scenarios: Competitive versus Cooperative

BACKGROUND

After decades of importing oil from Central Asia, the Near East, Sub-Saharan Africa, Western Europe, and the Western Hemisphere, China has employed motor vehicle fuel-saving technologies to sustain economic growth and development. These modern technologies consist of hybrid-electric vehicles (HEVs), compressed natural gas (CNG) vehicles and electric vehicles for public and private transportation. In this scenario the United States and other major oil consumers (Western Europe, Japan and India) adopt public and private transportation fuel-saving technologies and are importing considerably less oil in an effort to reduce their dependency on imported oil. It would not be realistic to expect China adopt transportation fuel-saving technologies if other large oil consumers did not. As a result of implementing fuel-saving technologies in motor vehicles for public and private transportation, China has cut its annual oil consumption by more than 50 percent. Consequently, global oil prices have dropped, oil consumption has been reduced, and the life of global oil supplies has been extended. The following two scenarios analyze energy futures in which China relies heavily on fuel-saving policies and technologies (hybrid, electric, compressed natural gas) to decrease consumption and increase production. In the competitive surplus scenario, competition (unilateral exploration) exists in the South China Sea for oil energy resources, while in the cooperative surplus scenario, cooperation (joint exploration) exists in the South China Sea for oil energy resources.

Competitive Surplus Scenario

Although China uses considerably less hydrocarbon-based fuels, China still explores for oil both onshore and offshore, including the East China Sea and South China Sea. Its unilateral oil exploration and production operations (upstream activities) in the East China Sea and South China Sea has caused considerable consternation with the governments of Brunei, Japan, Malaysia, the Philippines, Taiwan, and Vietnam because of disputed sovereignty claims.

Cooperative Surplus Scenario

This scenario is different from the competitive scenario in that China jointly explores for oil in the South China Sea and the Gulf of Mexico. Although China uses considerably less hydrocarbon-based fuels, it still explores for oil both onshore and off mainland China, including the South China Sea. China's efforts to conduct joint oil exploration and production operations (upstream activities) in the South China Sea have improved bilateral relations somewhat with Brunei, Malaysia, the Philippines, Taiwan, and Vietnam. China has also pursued joint oil exploration and production operations in the Gulf of Mexico with independent oil companies such as ExxonMobil and Chevron.

PROBABILITY OF OCCURRENCE

As in the competitive dependency scenarios analyzed in Chapters 2–8, in the surplus scenarios, a PRINCE score of greater than 60 percent indicates that future events/issues will occur or will likely continue if already in progress. A PRINCE score of 40 to 60 percent means that it is uncertain whether the events or issues will occur. Moreover, if any of the aforementioned issues or events has a PRINCE score of less than 40 percent, then they will not occur. See Appendix B for a description of the PRINCE Method. The overall probability of occurrence in the competitive surplus scenario is 77 percent, and the overall probability of occurrence for the cooperative surplus scenario is 54 percent, so the competitive surplus scenario is probable depending on the exact sequence of events and decisions taken, while the cooperative surplus scenario is uncertain.[1]

PRINCE scores were calibrated to improve forecasting accuracy by first averaging the scores of the China's fuel saving technology operations as one group and the scores for China's domestic crude oil and petroleum operations as one group. These scores were then averaged with the scores for each regional oil-prospecting operation in the South China Sea (Spratly Islands and the Paracel Islands) and each international oil-prospecting operation (either the East China Sea or Gulf of Mexico) to get the overall probability of occurrence for each scenario.

MACRO POLITICAL AND ECONOMIC STRATEGIC IMPLICATIONS FOR THE UNITED STATES

China's use of alternative fuel technology (hybrid electric vehicles, electric cars and compressed natural gas (CNG) cars in addition to gas-powered vehicles) in lieu of fossil fuel technology for public and private transportation will undoubtedly provide for better diplomatic relations with the United States with respect to global energy security and cooperation because China is no longer relying on primarily fossil fuels to sustain national economic growth and development. Essentially, China is not tapping in to the global oil market. The United States will not be energy insecure with respect to China because it faces far less competition from China in acquiring global oil energy resources. But because the United States has adopted fuel-saving technologies, like China, it also no longer depends on foreign oil. Petroleum cannot be used as a political weapon against the United States, so it is less entangled with unstable or hostile regimes, especially with those regimes that were major U.S. crude oil and petroleum producers in the past.

China's policy of using alternative fuel technologies will not reduce, change, or affect any global political issues between China and the United States that are unrelated to oil energy matters or energy security, such as international compliance of human rights, arms control (including weapons of mass destruction technology transfers), the tacit support of rogue regimes like Burma, Cuba, Iran, and Sudan, or the status of Taiwan. China will continue to build and improve its international political relationships with the developing and developed world using its non-ideological, pragmatic approach to maintain its domestic social stability, to ensure the political longevity of the Chinese Communist Party, to sustain its domestic economic growth and development, to increase its foreign trade, and to improve its global political position.

Macro Political and Economic Implications: Competitive Surplus Scenario

China's unilateral exploration, discovery, extraction, and refinement of crude oil in Japan's Exclusive Economic Zone (EEZ), in the Paracel Islands, and in the Spratly Islands may make the East China Sea and South China Sea politically unstable and disrupt the flow of international shipping traffic (including the transit of oil tankers) in the region. This is due in part to greater Chinese military presence in the East China Sea and South China Sea to protect its claimed oil interests in the Paracel Islands and in the Spratly Islands. Second, there is an increased probability of armed conflict between China, Japan, Brunei, Malaysia, Philippines, Taiwan, and Vietnam and possibly the United States over conflicting sovereignty claims in Japan's Exclusive Economic Zone and in the Paracel and Spratly Islands. This would most likely lead to a severe economic downturn, increasing the probability of political instability and the specter of war in the East China Sea and the South China Sea, resulting from the disruption of crude oil and petroleum exports to the United States. Increasing brinksmanship between the United States and China may further exacerbate regional tensions.

Macro Political, Economic, and Military Implications: Cooperative Surplus Scenario

Unlike the competitive surplus scenario, China's cooperative efforts to conduct joint exploration, discovery, extraction, and refinement of crude oil in the Paracel Islands and in the Spratly Islands may make the South China Sea more politically stable and may improve the flow of international shipping traffic (including the transit of oil tankers) in the region in the absence of sovereignty disputes and a kinetic military exchange. As well, China's joint operations with independent oil companies in the Gulf of Mexico adjacent to Cuba will make for greater bilateral diplomatic relations between China and the United States provided that the U.S. Congress has not disapproved any joint business agreements between the Chinese oil companies and the independent oil companies (namely U.S. oil companies), since oil is a strategic commodity and a national security issue.

China's operations in the South China Sea and East China Sea may make the region more economically stable if joint operations are successful and provide for better trade relations between China and ASEAN. China's cooperative efforts with independent oil companies in the Gulf of Mexico adjacent to Cuba will not necessarily make for greater bilateral economic and trade relations between China and the United States, but they may lead to greater bilateral trust and cooperation. However, this may be entirely dependent on the U.S. Congress' normalizing relations with Cuba.

MACRO-ECONOMIC IMPLICATIONS

China's use of alternative fuel technology (hybrid-electric vehicles, electric cars and compressed natural gas (CNG) cars in addition to gas-powered vehicles) in lieu of fossil fuel technology for public and private transportation will undoubtedly provide for better economic and trade relations with the United States, especially in motor vehicle sales, because of sharing and adoption of motor vehicle fuel-saving technologies between the United States, China, and Japan. The development and effective use of fuel-saving technologies in China will lower global oil prices, reduce oil consumption, and extend the life of available oil reserves. However, the success of fuel-saving technologies in China will largely depend on China's actual GDP growth rate and whether the international automotive manufacturing industry can capture China's domestic car sales market. Table 10.1 provides an estimate of GDP growth versus the number of motor vehicles in China in the near future, and Table 10.2 provides fuel economy comparisons for popular vehicles in 2009. Consumers could cut up to one-third of their annual fuel cost by driving fuel-saving technology vehicles.

China will not be deterred from supporting pariah states such as Burma, Cuba, Iran, and Sudan, prevent it from complying with international security and stability protocols on human rights and arms control (including weapons of mass destruction technology transfers). Beijing will continue to build and improve its international economic relationships (like its international political relationships) with the developing and developed world using its non-

Table 10.1 National Vehicle Fleet Projections for 3 GDP Growth Rates, China.

GDP Growth	*Year*	*Cars (millions of vehicles)*	*Motor Vehicles (millions of vehicles)*
10 percent	2010	13.9	42.5
10 percent	2115	24.4	68.4
10 percent	2020	43.1	110.2
8 percent	2010	11.4	36.0
8 percent	2115	18.0	52.9
8 percent	2020	28.5	77.8
6 percent	2010	9.3	30.4
6 percent	2115	13.2	40.7
6 percent	2020	18.7	54.5

Source: Calculated by G. Ingram based on data in China State Statistical Bureau (2002).

Table 10.2 Fuel Economy Comparisons for Popular Vehicles in 2009 based on 15,000 miles of annual traveling and a fuel cost of $2.44 per gallon

Make (2009)	*Model*	*MPG-City*	*MPG-Highway*	*Annual Fuel Cost*
Civic	Honda	25	36	$1,263
Civic-Hybrid	Honda	40	45	$871
Civic-CNG	Honda	24	36	$873
Accord	Honda	21	30	$1,526
Camry	Toyota	21	31	$1,464
Camry-Hybrid	Toyota	33	34	$1,076
Prius	Toyota	48	45	$794

(Source: Fuel Economy Guide 2009, www.fueleconomy.gov)

ideological, pragmatic approach to maintain its domestic and international objectives.

MACRO MILITARY IMPLICATIONS

China's use of alternative fuel technology will not necessarily make for better military relations with the United States in spite of potentially lower global oil prices, greater diplomatic opportunities, and trade possibilities in commercial vehicle alternative fuel technology between China, Japan, and the United States because of the lack of transparency regarding China's military capabilities and modernization and China's continued deployment of military personnel around in the world in support of its "go-out strategy." The United States will not necessarily be more energy secure because of China's military secrecy, modernization of its military forces, and growing military relations with the developing and developed world. Beijing will continue to support rogue states like Burma, Cuba, Iran, and Sudan and will continue to build and improve its international military relationships with them.

REGIONAL ANALYSES

In the surplus scenarios, analyses of China's oil energy activities and its implications for the United States are made for East Asia and Southeast Asia. Further, the Gulf of Mexico is individually addressed in the cooperative surplus scenario. Realistically, the impact of China's oil energy activities on the United States will vary (positively, negatively, or none) based on the extent of overall U.S. and Chinese political, economic/trade and military relations, aid and influence, and amount of oil energy resources available in the region. (See Appendix A for available oil resources.) Available hydrocarbon resources, proven crude oil reserves as of January 1, 2009 are shown below (see Table 10.3) according to the *Oil and Gas Journal.*

Although the Near East and the Western Hemisphere hold the majority of proven oil reserves and are clearly the most important regions of the world in terms of energy security, these regions are not vital to China's energy security because China uses fuel-saving technologies to sustain economic growth and development.

EAST ASIA IN THE COMPETITIVE AND COOPERATIVE SURPLUS FUTURES

The PRINCE model predicts that China will continue to prospect for oil in Japan's Exclusive Economic Zone by 71.4 percent China will continue to prospect for oil both onshore in the Tarim Basin by 91.4 percent, in Bohai Bay by 96.8 percent, and coastally by 97.5 percent. China will also develop and use fuel-saving technology vehicles by 96.7 percent.[2]

Table 10.3 Proven Oil Reserves

Region	*Billion Barrels*	*% of World Total*
Russia	60	4.47%
Asia-Pacific	34	2.53%
Central Eurasia	38.88	2.90%
Near East (Middle East & North Africa)	805.98	60.00%
Africa (Sub-Saharan Africa)	57.07	4.25%
Western Hemisphere (North & South America)	332.59	24.78%
Europe	13.66	1.02%
Grand Total	1,342.07	100%

Source: Oil and Gas Journal, 2009.

East Asia: Political, Economic, and Military Implications for the United States

China's oil energy activities in East Asia are unlikely to shift the balance of political power from the United States to China because of the strong diplomatic relationships that exist between the United States and Japan, nor does oil energy trading shape or influence political relationships in East Asia. Sino-Japanese diplomatic relations are tepid, and conversely U.S.-Japanese diplomatic relations are hot.[3]

The efficient development and effective use of fuel-saving technologies in China's motor vehicles will improve Sino-American energy security measures because China will no longer rely solely on fossil fuel energy resources to meet its domestic oil energy needs. In this sense, there would be no need for China to import as well as explore for and produce foreign oil (apart from the East China Sea and the South China Sea) because Beijing will have implemented public policies to transform both the public and private transportation industry using fuel-saving technologies. Although China will not have to compete with the United States for global oil energy resources, bilateral relations outside of energy security matters may not improve over current conditions. Contentious domestic and foreign policy issues like human rights violations and WMD technology transfers will remain.

China's hydrocarbon energy activities in East Asia are unlikely to shift the balance of economic power from the United States to China because of the strong economic relationships that exist (aside from oil energy trading) between the United States, China, and Japan. In this respect, oil energy trading does not play a pivotal role in shaping or influencing economic relations between the United States, China, and Japan. International energy trade in East Asia enhances regional stability as well as political and military relations, although bilateral trade relations in other commodities remain unaffected. Sino-Japanese and U.S.-Japanese economic relations are hot.[4]

Domestically, the adoption and use of fuel-saving technologies will result in greater fuel economy within China as well as a reduction in annual fuel cost for individual consumers. Table 10.2 provides a comparison of popular vehicle makes and models for the year 2009 to show how overall fuel economy will improve and how annual fuel costs (per unit) will be reduced when using fuel-saving technologies in motor vehicles.

China's hydrocarbon energy activities in East Asia are unlikely to shift the balance of military power from the United States to China or lead to a military competition for energy resources because of the close military relationship that exists between the United States and Japan. In this respect, hydrocarbon energy trading does not play a role in shaping or influencing military relations between the United States, China, and Japan and may not improve bilateral military relations in other areas. Sino-Japanese military relations are cool, and U.S. military relations with Japan are hot.[5] China's military modernization, its development of a blue-water navy, and increased global military-to-military ties will occur independently of the development and use of fuel-saving technologies in the public and private transportation sectors.

The Political, Economic, and Military Implications of China's Domestic Hydrocarbon Operations

Domestic hydrocarbon operations (exploration, discovery, extraction, and refinement) in territorial China such as the Tarim Basin in Xinjiang, Bohai Bay, and coastal China are unlikely to have any negative diplomatic, economic, or military consequences for the United States. These operations are sovereign in nature and do not involve the United States or any other country except in a capacity where technologically advanced countries like the United States may be able to provide public and private technical expertise to assist China in producing its natural hydrocarbon energy resources from underground and the below the ocean floor. Should Beijing seek greater cooperation from the United States, both countries will benefit on many levels, such as increasing their strategic petroleum reserves (SPR) as well as lowering global petroleum prices? Therefore it behooves both China and the United States to engage each other and work constructively.

SOUTHEAST ASIA IN A COMPETITIVE SURPLUS FUTURE

The PRINCE model predicts that China will continue to unilaterally prospect for oil in the Paracel Islands by 67 percent. China will continue to unilaterally prospect for oil in the Spratly Islands by 58 percent.[6]

The Political, Economic, and Military Implications of South China Sea Sovereign Dispute

There is a 67 percent chance that China will continue to extract oil from the Paracel Islands and a 58 percent chance that it will continue operations in the Spratly Islands.[7] Disputed sovereignty claims in the South China Sea over the Paracel Islands (between China, Taiwan, and Vietnam) and Spratly Islands (between China, Malaysia, the Philippines, Taiwan, and Vietnam) could draw the United States inexorably into an accidental war with China in the absence of preventive diplomatic measures. War would not only disrupt trade and commerce throughout Southeast Asia but the Asian Pacific as well, which could have first- and second-order effects on world trade. China's economy could potentially be undermined, leading to major social unrest and resulting in the overthrow of the Chinese Communist Party.

SOUTHEAST ASIA IN A COOPERATIVE SURPLUS FUTURE

The PRINCE model predicts that China will not jointly prospect for oil in the Paracel Islands with Taiwan and Vietnam. The model also predicts that China will not jointly prospect for oil in the Spratly Islands with Brunei, Malaysia, the Philippines, Taiwan, and Vietnam.[8]

The Political, Economic and Military Implications of Joint Exploration and Production in the South China Sea

The PRINCE scores indicate that under a cooperative scenario, China will not engage in joint oil exploration in the Paracel and Spratly Islands. If China and the nations of Southeast Asia (Malaysia, the Philippines, Taiwan, and Vietnam) work together in the South China Sea in joint oil exploration and production projects, there would be no need for the United States to become involved. Not only would such a scenario avert war, but shipping lanes would be far safer. In addition, China may be able to make up for lost oil energy imports imposed by this energy future. Overall, joint oil exploration and production in the South China Sea would not result in a political, economic, or military competition for oil energy resources between the United States and China.

GULF OF MEXICO IN A COOPERATIVE SURPLUS FUTURE

The PRINCE model predicts that China will jointly prospect for oil in the Gulf of Mexico with U.S. oil companies by 79 percent.[9]

The Political, Economic, and Military Implications of Joint Exploration and Production in the Gulf of Mexico with Independent U.S. Oil Companies)

Although it is impractical for China to explore for and produce oil beyond its shores, in this scenario there is a 79 percent chance that Beijing will engage in joint exploration and extraction of crude oil with independent U.S. oil companies in the Gulf of Mexico. Chinese oil companies such as the China National Petroleum Corporation (CNPC) and the China National Chemical Corporation (Sinopec) and independent oil companies such as ExxonMobil and Chevron will work as business partners to explore, discover, extract, and refine crude oil in the region. As a precondition for this arrangement, the U.S. Congress must approve any joint business agreements between the Chinese oil companies and independent oil companies, since oil is a strategic commodity and a national security issue. Trust-building measures may have to be undertaken first however, which is desired by China but may not be favored by factions in the U.S. Congress. Should the United States permit joint oil exploration operations in its backyard, the potential is that greater political, economic, and military bilateral relations could develop between the two countries.

SUMMARY

The emergence of the surplus scenarios is likely to be a fairly protracted process, since the conversion to alternative fuels is likely to be uneven over time. In any event, the wide spread emergence of alternative fuels/technologies within the next 100 years might be ushered in by an increase in domestic violence in some of the rentier oil producers if it is accompanied by a drop in oil revenues, which these governments dispense as a form of patronage. Regardless, China and the United States will not be competing for scarce energy resources, making it less likely for a change in the international status quo. Global issues that are unrelated to energy security such as international compliance of human rights, arms control (including weapons of mass destruction technology transfers), the tacit support of rogue regimes like Burma, Cuba, Iran, and Sudan, or the status of Taiwan are unlikely to reduce the latent political tensions between China and the United States.

Notes

1. See www.princeandpanda.com/media/6d4c7d4a7bb40f5bffff8019ffffa126.pdf.
2. See www.princeandpanda.com/media/6d4c7d4a7bb40f5bffff8019ffffa126.pdf.
3. Congressional Research Service, Library of Congress, *China's Foreign Policy and "Soft Power" in South America, Asia, and Africa: A Study Prepared for the Committee on Foreign Relations United States Senate* (Washington, D.C.: U.S. Government Printing Office), April 2008, 52, 55.

4. Congressional Research Service, Library of Congress, *China's Foreign Policy and "Soft Power" in South America, Asia, and Africa: A Study Prepared for the Committee on Foreign Relations United States Senate,* 52, 55.

5. Congressional Research Service, Library of Congress, *China's Foreign Policy and "Soft Power" in South America, Asia, and Africa: A Study Prepared for the Committee on Foreign Relations United States Senate,* 52, 55.

6. See www.princeandpanda.com/media/6d4c7d4a7bb40f5bffff8019ffffa126.pdf.

7. See www.princeandpanda.com/media/6d4c7d4a7bb40f5bffff8019ffffa126.pdf.

8. See www.princeandpanda.com/media/6d4c7d4a7bb40f5bffff8019ffffa126.pdf.

9. See www.princeandpanda.com/media/6d4c7d4a7bb40f5bffff8019ffffa126.pdf.

Chapter 11

Conclusions

This book examined how China's oil energy needs will likely shape future Sino-American relations under conditions of dependency and non-dependency. Scenario analysis was used to study China's oil energy needs because it provided the best way to establish a quantitative causality between China's alternative oil energy futures or scenarios and their strategic implications for the United States. Three plausible alternative future oil energy futures were developed and analyzed, determining their associated events and strategic political, economic, and military implications for the United States.

Using the PRINCE model of risk analysis forecasting, it was determined empirically that the competitive dependence and competitive surplus energy futures have some probability of occurring because the average PRINCE score (of all the independent events and issues) for each scenario was greater than 60 percent. If a scenario had an average PRINCE score between 40 and 60 percent, then it was uncertain whether a scenario would occur. Moreover, if a scenario had an average PRINCE score (of all the independents events and issues) of less than 40 percent, then a scenario would not occur. It is uncertain as to whether the Cooperative Surplus energy future would occur because of a PRINCE score's falling between 40 percent and 60 percent. This information is useful because it tells us that China will not necessarily follow a path of oil energy dependence according to the status quo. The scenarios are illustrated graphically below in Figure 11.1 on a continuum based on their probability of support and likelihood of occurrence and in Table 11.1. Note

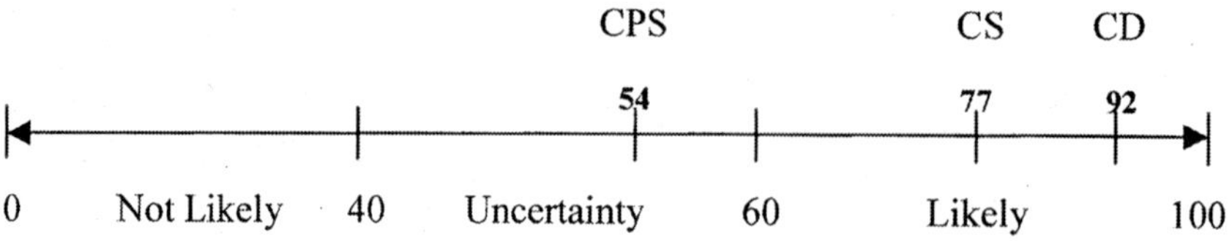

Competitive Dependency (CD)
Competitive Surplus (CS)
Cooperative Surplus (CPS)

Figure 11.1 Probability of Occurrence for Three Chinese Oil Energy Futures

that the cooperative dependency energy future was not considered because it was an unrealistic scenario. See Chapter 1 for a discussion on why this oil energy future is not feasible.

The competitive dependency oil energy future has the highest probability of support average—that is, *92* percent—based on the average PRINCE scores of 71 independent events or issues followed by the competitive surplus (non-dependency) future with probability of support average of 77 percent, based on the calibrated average PRINCE scores of nine independent events or issues and the cooperative surplus (non-dependency) future with a probability of support average of 54 percent, based on the calibrated average PRINCE scores of nine independent events or issues. Even though the competitive dependency energy future has the highest probability of support average of all three alternate energy futures and appears to be the most likely energy future, the competitive surplus energy future is also probable. See Table 11.1.

This research shows that the United States and China will most likely compete for scarce oil energy resources under oil dependency conditions and are less likely to use fuel-saving technologies competitively or cooperatively under oil nondependency conditions. Obviously, competition between the United States and China for scarce oil energy resources will also generate strategic consequences for poor but resource-rich producer countries that will still be afflicted with the "resource curse," and oil could still be used as a political weapon against the United States and China. What is most troubling is that China could find itself making political concessions such as arms-for-oil deals. China's energy insecurity could fuel global insecurity if it were to exchange weapons of mass destruction technologies for reliable oil supplies, as in the case of Iran.[1]

Through the adroit use of soft power, China's strategic interactions with oil supplier states of the non-free and partly-free regions of the world provides

Table 11.1 Chinese Oil Energy Futures or Scenarios

Oil Energy Future/Scenario #1: ***Competitive Dependency*** (Oil Dependency) ***Highest Probability of Support (92%)***	Oil Energy Future/Scenario #2: ***Competitive Surplus*** (Oil Non-Dependency)
Oil Energy Future/Scenario #4: ***Cooperative Dependency*** (Oil Dependency) ***This future/scenario was not considered.***	Oil Energy Future/Scenario #3: ***Cooperative Surplus*** (Oil Non-Dependency)

them with an alternative political and economic model, a model of rapid-state-sponsored economic growth and modernization alongside political authoritarianism, which undermines U.S. efforts to promote democratization, good governance, and transparency reforms as in the case of Angola and Nigeria. China's pragmatic, non-ideological approach to bilateral relations provides foreign governments with an alternative partner in international affairs, thereby reducing U.S. influence.[2] In this respect, China, does not link policy issues to commercial oil ventures, making it more attractive as a trading partner and strengthening its global influence. U.S. international leverage, such as the ability to garner political support in the United Nations to enforce economic sanctions against rogue trading states like Iran and Sudan, is weakened. Overall, U.S. dependence on foreign oil in competition with China lessens its international leverage, undermines its foreign policy initiatives, and provides an incentive for Washington to continue entanglement with hostile and/or unstable regimes.[3]

Economically, China has the potential to further undermine U.S. efforts to acquire cheap oil, especially when it is able to use lucrative infrastructure investments to entice oil suppliers to favor China's national oil companies over independent oil companies in exploration and production contracts as in Angola. Note that China is Angola's largest trading partner. Consequently, energy trading blocs favor China's national oil companies over the U.S.-based independent oil companies (IOCs) operating in politically nonaligned countries in the developing world. Unless China can find a way to reduce its oil dependence, its hard power and strategic military engagement will grow to ensure its energy security. China has been accelerating its own submarine development and acquiring more Russian submarines capable of operating long distances. China is also contemplating the addition of a nuclear aircraft carrier (CVN) to its fleet in the near future. China has also been quietly engaged in a strategy of military and economic cooperation with Pakistan, Myanmar, Bangladesh, Sri Lanka, and the Maldives in the Indian Ocean region in order to build and/or have access to ports and airbases to support not only a future tanker fleet but also a blue-water navy capable of projecting power not only in the Indian Ocean region but also in the Persian Gulf.

Chinese warships are currently patrolling the Bab-el-Mandeb in support of international anti-piracy operations. Beyond the Indian Ocean region, China established an intelligence collection facility in Cuba and has been offered the use of a former U.S. airbase in Ecuador. China is also engaged in a weapons-for-oil exchange with Iran. It is Iran's largest conventional weapons supplier and its greatest supporter of nuclear technology development. China coordinates its overseas military operations in support of the larger foreign, diplomatic, economic, and security agenda established by its leadership. Although there are few conditions for a military confrontation over oil energy resources, it is not inconceivable that the Chinese Navy may be able to project sea power and establish sea control well beyond its shores, including the sea lines of communication (SLOCs) to ensure its energy security in the near future.

China's rising dependence on imported oil, adroit soft power, economic prowess, strategic engagement of the world as an alternative model of political and economic development, military modernization, and Mahanian naval desirability in tandem with rapid economic growth as the fastest growing economy in the world in 30 years with an average GDP growth rate above 10 percent can only mean that it will alter the global status quo and become the dominant actor in world affairs.[4] The United States will be less influential politically, economically, and militarily because China is skillfully harnessing and strategically exercising the elements of national power (diplomatic, informational, military, and economic) to acquire scarce oil energy resources in the Near East, the Western Hemisphere, and Sub-Saharan Africa and is becoming a great power as a consequence.

Whether the United States will be a satisfied defender when this change occurs is uncertain. What is certain is that China will replace the United States as the dominant power in the world unless the United States can find a way to be less dependent on foreign oil. One possible way is through alternative fuels/technologies. However, the emergence of alternative fuels/technologies future is likely to be a fairly protracted process, since the conversion to alternative fuels is likely to be uneven over time. Regardless, the widespread emergence of alternative fuels/technologies might be ushered in by an increase in domestic violence in some of the rentier oil producers if it is accompanied by a drop in oil revenues, which these governments dispense as a form of patronage.

The question that emerges from this exhaustive research on China's energy insecurity is: What kind of power will it be if it alters the status quo? Will it be a responsible stakeholder of the international system, or will it contribute to global insecurity? These are questions that policymakers and government officials in China and United States must ponder with respect to China's dependence on foreign oil.

U.S. POLICY RESPONSES FOR THE COMPETITIVE DEPENDENCY SCENARIO

The competitive dependency scenario will have the greatest political, economic, and military implications for the United States if this scenario should materialize. Therefore, the United States should act accordingly by implementing unilateral, bilateral, and multilateral policies that are consistent with its national security interests. However, any U.S. policy initiative should be judicious in nature and cautious in approach and should not diminish or damage Sino-American diplomatic ties, economic, and trade relations or global trade. Otherwise, the consequences will be grave for global political and economic stability as well as Sino-American relations.

In conjunction with its public policy, the United States should adopt a sophisticated strategy that utilizes the full array of diplomatic, information, economic, and military tools available in a reasonably coordinated fashion to influence China's energy outcomes.[5] The goals such a strategy should be:[6]

- To build the mutual trust necessary for effective cooperation
- To increase China's willingness to work with its United States counterparts to develop and deploy capabilities to address China's oil energy problems
- To enhance incentives and opportunities for China to follow a desired energy strategy
- To help China develop specific technical and policy capabilities that will better enable it to achieve desirable energy outcomes

Foremost for any U.S. policy with respect to China is trust. According to Lieberthal and Herberg, "trust is both fundamental and nettlesome."[7] China should trust the United States enough to allow effective cooperation on any energy matters. Trust becomes important in reducing China's acute security concerns regarding the sourcing and transporting its imported oil energy.[8] However, China's oil sourcing strategy is rooted in distrust of U.S. strategic intentions towards its energy security requirements.[9] A case and point would be CNOOC's failed bid for Unocal in the summer of 2005. Moreover, China's state equity oil policy and the tough global competitive environment for China's state-owned oil companies have driven it into deals with problem states like Iran and Sudan.[10] These deals in turn have produced outcomes that have further eroded trust between the United States and China.[11] Overall, trust between the United States and China should be established to provide for more effective cooperation on oil energy matters. Unfortunately, energy cooperation is inevitably hostage to the broad sweep of Sino-American relations.[12] However, a specific level of bilateral and multilateral cooperation in the energy world can play a significant role in either enhancing or reducing overall mutual trust.[13] It behooves the United States and China to have a greater awareness of this reality.[14]

RECOMMENDATIONS FOR BILATERAL INITIATIVES IN ENERGY COOPERATION

The United States and China have already undertaken a substantial number of bilateral cooperative efforts in the energy domain under the aegis of the Fossil Energy Protocol. However, a high level of engagement is critical because no sector is more intimately tied to the fate of the global economy than energy, particularly oil energy. To this extent, the United States should:

- Convene bilateral dialogues on energy cooperation to build trust;
- Promote cooperative dialogues on oil-saving technologies (particularly transportation technology);
- Encourage China to take on a greater role within global economic, trade, and financial institutions and mechanisms, including the Organization for Economic Cooperation and Development Export Credit Arrangement, International Energy Agency (IEA), World Bank and G-8 process; and
- Encourage greater military-to-military ties with China, especially regarding the observation and participation in peacetime military training, exercises, and operations related to the area of energy security.

Each of initiatives listed above is self-explanatory, but it is helpful to explain the role and efficacy of each initiative with respect to bilateral energy cooperation.

Bilateral Dialogues (Phase One). Bilateral dialogues between the United States and China would be an important first step to build trust in energy cooperation. Dialogues would be non-statutory in nature, conducted formally or informally, and based on shared concerns on energy issues affecting both countries such as demand, consumption and security.

Cooperative Dialogues on Oil-Saving Technologies (Phase Two). In addition to promoting bilateral dialogues, the United States can also promote specific dialogues on oil-saving technologies as the next step in energy cooperation between the United States and China. Dialogues should promote the development and adoption of plug-in hybrid engines, the sustainable production of biofuels, and fuel efficiency improvements, with the shared goal of reducing demand growth of fossil fuels.

Bilateral Military Cooperation. Military cooperation especially naval cooperation in the area of energy security, would help to build trust between China and the United States, since China does not have the ability to protect the SLOCs militarily in the event of a supply disruption caused by piracy, terrorism, or naval blockade. Typically, the security of the SLOCs falls upon the United States Navy and its allies in the form of Combined Task Forces (CTFs) operating in the Gulf of Aden, the Gulf of Oman, the Arabian Sea, the Indian Ocean, off the Somali Coast, and in the Arabian Gulf. If the United States could encourage China to operate jointly with one or more of the combined task forces rather than independently, then trust could be built between the United States and China and alleviate some concerns that China may have with respect to potential supply disruptions by U.S. naval forces. Further, China's naval participation would help to reduce U.S. expenditures to defend oil supplies from the Arabian Gulf.

OECD Export Credit Arrangement. This is a special-purpose regime composed of nine participants (Australia, Canada, the European Community, Japan, South Korea, Norway, New Zealand, Switzerland, and the United States) with two basic objectives: to control the predatory use of export finance to shift rents from one supplier state to another and to control the unintentional transfers of wealth from supplier to buyer states.[15] It sets and enforces terms and conditions on official export credits and foreign aid.[16] Further, it plays an important role in the multilateral trading system to provide a level playing field so that both OECD and non-OECD exporters compete on the price and quality of their goods and services, not on the support they receive from their governments.[17]

The success of the Export Credit Arrangement derives from its participants' aims to eliminate trade distortion and level the playing field, driven by economic, competitive, and developmental motivations and the need to reach practical and pragmatic consensus.[18] Although China is not a member in the Export Credit Arrangement, membership would help it become a responsible stakeholder in the international system by conforming to global arrangements governing state-backed finance because state-backed finance can distort energy markets.[19] Thus, China's use of predatory financing would be diminished and would not undermine an open and competitive world oil market.

RECOMMENDATIONS FOR MULTILATERAL INITIATIVES IN ENERGY COOPERATION

Multilateral initiatives offer a very important basis for building trust in the energy world in light of the many problems in Sino-American bilateral relations. Therefore, it behooves the United States to consider and promote the following initiatives because each has the potential to produce positive outcomes in energy cooperation and to reduce Chinese mistrust of long-term U.S. intentions in the energy world. To this extent, the United States should:

- Convene multilateral dialogues on energy cooperation to build trust;
- Encourage international partnerships between national oil companies and independent oil companies that demonstrate the win-win potential of international cooperation in oil investment;
- Broaden diplomatic dialogue inside regions where there are significant oil energy investments to identify overlapping interests, such as stable investment conditions, and possible areas for cooperation, including foreign assistance; and
- Persuade global economic, trade, and financial institutions such as the G-8 and International Energy Agency (IEA) to invite China as a member to show how the international system can directly benefit China's energy security.

Each of initiatives listed above is self-explanatory, but it is helpful to explain them further in terms of their role and efficacy in terms of multilateral energy cooperation.

Multilateral Dialogues. Multilateral dialogues conducted in international forums like the United Nations and the World Trade Organization would be an important initial step to building trust in energy cooperation between the United States and China. The advantage of conducting multilateral dialogues in international forums like the United Nations and the World Trade Organization as opposed to the G-8 and the International Energy Agency is that China is already a member of the two former organizations. Membership in the two latter organizations requires China to become an advanced industrialized economy in the case of the G-8 or a member of the Organization for Economic Cooperation and Development (OECD) in the case of the International Energy Agency. To qualify for membership in the OECD, China must have a per capita GDP of at least $3,000 and meet certain standards of human rights. Therefore, "any effort to relax existing conditions to accommodate Chinese membership would prove to be too divisive among current IEA members."[20]

International Partnerships. Encouraging international partnerships between national oil companies and international oil companies such as the Sinopec-Saudi Aramco-ExxonMobil Fujian refinery upgrade project not only demonstrates the win-win potential of international cooperation but also helps to solidify the rules of the game with respect to oil energy investments.

Regional Diplomatic Dialogues. Broadening diplomatic dialogues in oil-rich regions to identify overlapping interests, such as stable investment conditions, and possible areas for cooperation, including foreign assistance with respect to oil energy investments, can bring about a reduction in conflict and build trust and cooperation between the United States, China, and oil energy producers.

The G-8. The G-8 is an informal forum representing the most advanced industrialized democracies that meets annually to address major economic issues of common concern. The G-8 would benefit from revising its institutional mission by becoming a forum that includes not just only the advanced industrial democratic states but also the most consequential countries in the world to address the most pressing economic issues, such as energy.[21] The United States can play an important role by persuading the G-8 to invite China as well as encouraging China to participate in the G-8 and thereby become a major global stakeholder. It does not make sense to exclude China from an international regime like the G-8 if such exclusion encourages China to free ride on the benefits of that regime without bearing any serious responsibility for its contributions or enforcement.[22]

The IEA. The International Energy Agency is a Paris-based intergovernmental organization established under the OECD in the wake of the 1974 oil crisis. It was initially focused on formulating multilateral measures that would reduce the vulnerability of its member countries to future oil supply disruptions but has since broadened its focus on energy policy in the areas of energy security, economic development, and environmental protection (the International Energy Agency).

Although China has participated in IEA workshops and related activities, China is not a member of the IEA because it is not a member of OECD. If the United States could persuade the IEA to make China an actual IEA member with full participation in all activities and full assumption of responsibilities for the management of supply disruptions, then the United States would be sending a strong signal to China that it is serious about treating China as an important player in the international energy regime.[23]

RECOMMENDATIONS FOR UNILATERAL INITIATIVES IN ENERGY COOPERATION

The United States consumes about 25 percent of all the oil produced in the world, yet it accounts for 10 percent of world oil production.[24] Moreover, the

United States imports over 50 percent of what it consumes. Importing oil imposes costs affecting the national security of the United States, including the dedication of military assets to protect overseas oil interests on land and at sea. Therefore, it behooves the United States to consider public policies that effectively would mitigate disruptions in the supply of oil, expand domestic sources of supply, and reduce domestic consumption of oil. Such policies should include the following initiatives:

- Open Access to environmentally sensitive and other restricted areas;
- Draw on the Strategic Petroleum Reserve (SPR);
- Raise Corporate Average Fuel Economy (CAFÉ) Standards;
- Promote federal purchasing of plug-in hybrid electric vehicles;
- Provide tax credits to purchasers of plug-in hybrid electric vehicles;
- Encourage the private business and philanthropic sectors to support more in-depth research and analysis of energy security; and
- Create an APRA-E modeled after DARPA for federal research into clean energy technologies.

Each of the initiatives listed above is self-explanatory, but it is helpful to explain them further in terms of their benefits to United States energy security.

Open Access to Environmentally Sensitive and Other Restricted Areas. Oil companies have been restricted from drilling in some environmentally sensitive areas in the United States. Two such areas are the Arctic National Wildlife Refuge (ANWR) in Alaska and the Outer Continental Shelf (OCS) off both the east and west coasts of the United States. Expanded access to these areas might add to the global oil supply an amount equal to 4 percent to 11 percent and generate approximately $161 billion in tax revenue for state and federal governments.[25]

Drawing on the Strategic Petroleum Reserve. The Strategic Petroleum Reserve (SPR) was established in 1975 to mitigate supply disruptions in times of national emergency. As of August 2008, the SPR held more than 700 million barrels of oil, equivalent to nearly 60 days of imported oil.[26] However, a large-scale drawdown of the SPR has not occurred. U.S. policy for use of the SPR is ambiguous, reducing its efficacy.[27] Policymakers have been reluctant to state what would trigger SPR use, since under current law, this would mean defining in advance what constitutes a national emergency.[28] If the United States stated publicly that the SPR will be definitely used in the event of a major supply disruption, the oil market would be better informed and likely to act more temperately if such an event materialized.[29]

Raise Corporate Average Fuel Economy (CAFÉ) Standards. Corporate Average Fuel Economy (CAFE) standards were first implemented in 1975 following the

1974–1975 Arab oil embargo and have been the primary policy used to promote improved car and light-truck fuel economy in the United States over the last three decades.[30] However, fuel economy improvements have been limited since the 1980s despite technological advances because fuel economy requirements differed between cars and light trucks. CAFE standards were less stringent for light trucks relative to cars to avoid penalizing users of work trucks.[31] Further, light trucks, which included sports utility vehicles (SUV) and minivans, increased their market share in automobile sales relative to cars, which resulted in lower fuel economy improvement. Thus, light truck lower fuel economy standards and increased domestic sales contributed to limited overall fuel economy improvements in the last 20 years.[32] If available technology were applied and CAFE standards were updated on a regular basis, car and light truck fuel economy standards would improve. and U.S. oil demand would be lowered by 3 million–5 million barrels a day by 2030.[33]

Federal purchasing of plug-in hybrid electric vehicles (PHEVs). Since the federal government purchases more than 65,000 new cars each year, it could potentially change the automobile industry by ordering 30,000 plug-in hybrid vehicles for delivery within the next two-years.[34] This would help to jump start the market for PHEVs, help finance conversion of existing production lines, and create economies of scale.[35]

Provide tax credits to purchasers of plug-in hybrid electric vehicles. Tax credits to purchasers of plug-in hybrid electric vehicles could play a pivotal role in transforming the U.S. vehicle fleet.[36] For example, a commitment to provide $6,000 in tax credits to purchasers of the first million fuel plug-in hybrids would accelerate the deployment of such vehicles.[37] Moreover, tax credits to purchasers of other fuel-efficient cars and trucks would also accelerate the deployment of such vehicles.[38] In conjunction with policies to assist automakers to retool production lines, tax credits could help revitalize the U.S. auto industry.[39]

Encourage the private business and philanthropic sectors to support more in-depth research and analysis of energy security. The U.S. government generally supports research and analysis of energy security as well as energy-saving technologies by nonprofit public policy organizations like RAND, the Brookings Institution, and the Center for Strategic and International Studies (CSIS). Although the private business and philanthropic sectors support some nonprofit energy-saving technology research and analysis, they generally do not support energy security research and analysis beyond proprietary research and analysis because energy security, for the most part, is the responsibility of the public sector, such as the U.S. government. Further, the public sector generally commissions and finances nonprofit public policy research and analysis. If the private business and philanthropic sectors could be encouraged to support in-depth research and analysis as well as nonprofit public policy research and analysis of energy security, their efforts would add to the body of knowledge of what is known as well as what is not known about energy security because of the plethora of expertise available at their fingertips (such as risk analysts or energy economists). The private business and philanthropic sectors could also supplement federal research funding for energy security research and analysis.

Create an APRA-E modeled after DARPA for federal research into clean energy technologies. The National Academy of Sciences recommended that a new federal energy research agency, "APRA-E," be created and modeled after the Defense Advanced Research Projects Agency (DARPA) with an initial annual funding of $300 million.[40] If successful, increased federal research funding could be committed to developing alternatives to oil.

RECOMMENDATIONS FOR REGIONAL INITIATIVES

In this energy future scenario, regional initiatives are proposed for the Asia-Pacific, Central Eurasia, the Near East (Middle East and North Africa), Sub-Saharan Africa (Africa), Western Europe, and the Western Hemisphere. The initiatives will vary in size and scope in light of overall U.S. and Chinese strategic engagement and available oil energy reserves in each region.

ASIA (INCLUDES OCEANIA, PAKISTAN, RUSSIA, AND THE SOUTH CHINA SEA)

Asia is one of the least energy-rich regions of the world, with approximately 7 percent of the world's proven oil reserves. Nevertheless, Asia is vital to China's strategic interests because most of China's imported crude oil and petroleum is transported by ships transiting through the Straits of Malacca and the South China Sea. Given the lack of trust between the United States and China with respect to energy security and the potential for accidental military escalation in the East China Sea and South China Sea, it would behoove the United States to undertake key unilateral, bilateral, and multilateral initiatives in the region not only to build trust between the United States and China but to minimize the potential for accidental war. Such initiatives that the United States should consider would include:

- Create a perception that the United States is engaged in Asian affairs and not just the Global War on Terror;
- Involve China in joint naval efforts to patrol the Straits of Malacca to contribute to the stability of sea lines of communications;
- Create patterns of cooperation with China on regional security issues such as energy security so China's growing role is in the direction of cooperation rather than conflict;

- Ensure regular, senior-level contact with Chinese and Asian leaders to anticipate and preempt potential areas of conflict and disagreement, especially with respect to the East China Sea and the South China Sea; and
- Empower regional institutions, such as the ASEAN Regional Forum (ARF), through forms of multilateral cooperation to address energy security challenges.

Each of initiatives listed above is self-explanatory, but it is helpful to explain them further in terms of their benefits to United States energy security.

Create a perception that the United States is engaged in Asian affairs and not just the Global War on Terror. The Global War on Terror (GWOT) dominated the Bush Administration's foreign policy agenda in the wake of the events of September 11, 2001. While combating international terrorism made America feel safe again domestically, it hurt America's international relations. The United States would benefit regionally by downplaying the U.S.-led Global War on Terror and placing more emphasis on its normal relations, especially since there is a perception that America is only interested in fighting Islamic militants.

Involve China in joint naval efforts to patrol the Straits of Malacca to contribute to the stability of sea lines of communications. Naval cooperation in the area of energy security would help to build trust between China and the United States since China does not have the ability to protect the SLOCs militarily in the event of a supply disruption caused by piracy, terrorism, or a naval blockade. Typically, the security of the SLOCs falls upon the United States Navy and its regional allies. If the United States could encourage China to operate jointly, then trust could be built between the United States and China and alleviate some concerns that China may have with respect to potential supply disruptions by U.S. naval forces.

Create patterns of cooperation with China on regional security issues such as energy security so China's growing role is in the direction of cooperation rather than conflict. Chinese as well as U.S. oil energy demand and consumption will increase over the next 20 years. Further, China's role on the international stage is expected to rise in conjunction with increased energy demand and consumption, which may or may not have adverse consequences for bilateral relations. It would be beneficial to the United States to find novel ways, both formally and informally, encourage China to participate in confidence-building measures (CBMs) to reduce tensions and the specter of an accidental war in the region.

Ensure regular, senior-level contact with Chinese and Asian leaders to anticipate and preempt potential areas of conflict and disagreement, especially with respect to the East China Sea and the South China Sea. Regular, senior-level contact with Chinese and Asian leaders, especially military leaders in the form of military-to-military relations, is vital to deconfliction, especially since the potential for accidental war is great.

Further, track-two dialogues (informal relations) can influence other actors in the security and defense establishments, such as naval officers, through military-to-military maritime dialogues.[41] In fact, track-two dialogues involving maritime confidence building has proved less contentious than other regional issues and has appeared to be a productive area for cooperative security practices in South Asia.[42]

Empower regional institutions, such as the ASEAN Regional Forum (ARF), through forms of multilateral cooperation to address energy security challenges. The ASEAN Regional Forum (ASF) has been regarded as a better model for security cooperation in non-Western regions than the European experience of highly institutionalized regional cooperation because of its emphasis on personal contacts, informality, and consensus-building rather than formal institutionalized decision-making.[43] Consisting of 27 participants, the ARF's objectives are to foster dialogue and consultation, and promote confidence building and preventive diplomacy in the region. If the United States can empower the ARF to address pressing energy security challenges in South East Asia, it may help to reduce the possibility of accidental war in the South China Sea and possibly the East China Sea.

CENTRAL EURASIA

Central Eurasia is not particularly energy rich with approximately 2.9 percent of the world's proven oil reserves. Nevertheless, Central Eurasia is important to China in terms of energy security because China has unfettered access to the region's crude oil reserves without interference from the United States. Given the lack of trust between the United States and China with respect to energy security, it would behoove the United States to undertake key multilateral initiatives to build bilateral trust in the region. Such initiatives that the United States should consider include:

- Create patterns of cooperation with China on regional security issues such as energy security so China's growing role is in the direction of cooperation rather than conflict; and
- Empower regional institutions, such as the Shanghai Cooperation Organization (SCO), through forms of multilateral cooperation to address energy security challenges.

Each of initiatives listed above is self-explanatory, but it is helpful to explain them further in terms of their benefits to United States energy security.

Create patterns of cooperation with China on regional security issues such as energy security so China's growing role is in the direction of cooperation rather than conflict. China's role on the international stage is expected to rise in conjunction with increased energy demand and consumption, which may or may not have adverse consequences for bilateral relations. It would be beneficial to the United States to find novel ways, both formally and informally, to encourage China to participate in confidence-building measures (CBMs) to reduce tensions and build trust.

Empower regional institutions, such as the Shanghai Cooperation Organization (SCO), through forms of multilateral cooperation to address energy security challenges. Although the Shanghai Cooperation Organization is primarily focused on its member nations' Central Asian security-related concerns like terrorism, separatism, and extremism, expanded dialogue regarding the potential energy security challenges facing the region would help to improve relations not only between members but also build trust. The United States could play an important role in this respect even though it is not an SCO member.

THE NEAR EAST (MIDDLE EAST AND NORTH AFRICA)

The Near East is the most energy-rich region with approximately 60 percent of the world's proven oil reserves. Further, the Near East produces almost 50 percent of China's crude oil and petroleum, compared to 23 percent for the United States. Because the region produces almost 50 percent of China's oil, it is critical to China's energy security, especially since the oil is transported by ships along the sea lines of communication from the Persian Gulf to the South China Sea and is subject to potential supply disruptions by pirates, terrorists, and the United States Navy. Given the lack of trust between the United States and China with respect to energy security, China's growing influence in the region with major oil producers like Iran and Saudi Arabia and the potential for accidental war between the United States and Iran, it would behoove the United States to undertake key unilateral, bilateral, and multilateral initiatives in region not only to build trust between the United States and China but to minimize the potential for accidental war. Such initiatives that the United States should consider include:

- Create a perception that the United States is engaged in Near East affairs and not just the Global War on Terror;
- Engage rogue oil states like Iran using track-two diplomacy to manage conflict;

- Involve China in joint naval efforts to patrol the Arabian Gulf to contribute to the stability of sea lines of communications;
- Encourage international partnerships between national oil companies and independent oil companies that demonstrate the win-win potential of international cooperation in oil investment in the Near East;
- Create patterns of cooperation with China on regional security issues such as energy security so China's growing role is in the direction of cooperation rather than conflict; and
- Ensure regular, senior-level contact with Chinese and Near East leaders to anticipate and preempt potential areas of conflict and disagreement, especially with respect to the possibility of accidental military conflict in the Arabian Gulf.

Each of the initiatives listed above is self-explanatory, but it is helpful to explain them further in terms of their benefits to United States energy security.

Create a perception that the United States is engaged in Asian affairs and not just the Global War on Terror. The Global War on Terror (GWOT) dominated the Bush Administration's foreign policy agenda in the wake of the events of September 11, 2001. While combating international terrorism made America feel safe again domestically, it hurt America's international relations. The United States would benefit regionally by downplaying the U.S.-led Global War on Terror and placing more emphasis on its normal relations, especially since there is a perception that America is only interested in fighting Islamic militants.

Engage rogue oil states like Iran using track two-diplomacy to manage conflict. Track-two dialogue, which is defined as "unofficial dialogue, focused on problem solving, in which the participants have some form of access to official policymaking circles," has had moderate success in managing conflict in the Near East. If the United States could engage Iran using track=two diplomacy, political tensions could be reduced, and trust could be built. However, the success or failure of track-two diplomacy depends on whether the participants are open-minded about change and have access to official policymaking circles. Otherwise, such efforts are futile.

Involve China in joint naval efforts to patrol the Arabian Gulf to contribute to the stability of sea lines of communications. Joint naval cooperation in the Arabian Gulf would help to reduce tensions and build trust, especially since the United States and its allies patrol the region. China's naval participation would help to reduce U.S. expenditures to defend oil supplies from the Arabian Gulf.

Encourage international partnerships between national oil companies and independent oil companies that demonstrate the win-win potential of international cooperation in oil investment in the Near East. Encouraging international partnerships between national oil companies and international oil companies in the Near East not only demonstrates the win-win potential of international cooperation but also helps to solidify the rules of the game with respect to oil energy investments.

Create patterns of cooperation with China on regional security issues such as energy security so China's growing role is in the direction of cooperation rather than conflict. China's role in the Near East is expected to rise in conjunction with increased energy demand and consumption, which may or may not have adverse consequences for U.S.-China bilateral relations. It would be beneficial to the United States to find novel ways, both formally and informally, to indulge China to participate in confidence-building measures (CBMs) to reduce tensions and build trust, especially since the Near East produces approximately 50 percent of China's oil.

Ensure regular, senior-level contact with Chinese and Near East leaders to anticipate and preempt potential areas of conflict and disagreement, especially with respect to the possibility of accidental military conflict in the Arabian Gulf. Regular, senior-level contact with Chinese and Near East leaders, especially military leaders in the form of military-to-military relations is vital to conflict reduction, especially since the potential for accidental war is great in the Arabian Gulf.

SUB-SAHARAN AFRICA (AFRICA)

Africa is one of the least energy-rich regions of the world with approximately 4.25 percent of the world's proven oil reserves. However, the region produces almost 30 percent of China's crude oil and petroleum compared to 14 percent for the United States. Because the region produces almost 30 percent of China's oil, it is critical to China's energy security, especially since oil is transported by ships along the sea lines of communication from key oil-producing states like Angola, Nigeria, and Sudan and is subject to potential supply disruptions by pirates, terrorists, and the U.S. Navy. Given the lack of trust between the United States and China with respect to energy security, China's growing influence with major oil producers like Angola, Nigeria, and Sudan, and the potential for supply disruptions, it would behoove the United States to undertake key unilateral, bilateral, and multilateral initiatives in the region not only to build trust between the United States and China. Such initiatives that the United States should consider include:

- Expand diplomatic presence in Africa;
- Encourage and support U.S. businesses (especially IOCs) to invest more in Africa;

- Create a perception that the United States is engaged in African affairs and not just the Global War on Terror;
- Engage rogue oil states like Sudan using track-two diplomacy to manage conflict;
- Involve China in joint naval efforts to patrol coastal Africa in addition to the Horn of Africa to contribute to the stability of sea lines of communications;
- Encourage international partnerships between national oil companies and independent oil companies that demonstrate the win-win potential of international cooperation in oil investment in Africa;
- Create patterns of cooperation with China on regional security issues such as energy security so China's growing role is in the direction of cooperation rather than conflict; and
- Ensure regular, senior-level contact with Chinese and African leaders to anticipate and preempt potential areas of conflict and disagreement.

Each of the initiatives listed above is self-explanatory, but it is helpful to explain them further in terms of their benefits to United States energy security.

Expand diplomatic presence in Africa. Compared to other regions of the world, Africa receives the least amount of U.S. political and diplomatic attention. It was not until after the events of September 11, 2001, that Africa merited further U.S. interest, particularly in terms of expanded military operations in the Global War on Terror and the creation of the United States African Command. African policymakers perceive the U.S. primary role in Africa to be military in nature and not political/diplomatic. The United States needs to reevaluate how it is viewed in Africa and focus more on in its diplomatic imperatives. The United States needs to expand its diplomatic presence in Africa and devote more attention to its political/diplomatic relations with individual African states.

Encourage and support U.S. businesses (especially IOCs) to invest more in Africa. U.S. businesses can be important sources of revenue to fund needed infrastructure projects in some of the poorest and underdeveloped areas of Africa in addition to public sector funding. Funding not only helps the impoverished and malnourished people of Africa, but it also helps to build political goodwill between nations. If the United States could encourage more U.S. businesses, especially IOCs, to invest more in Africa, it would certainly help the U.S. mandate to appear less militaristic.

Create a perception that the United States is engaged in African affairs and not just the Global War on Terror. The Global War on Terror (GWOT) dominated the Bush Administration's foreign policy agenda in the wake of the events of September 11, 2001. While combating international terrorism made America feel safe again domestically, it hurt America's international relations. The United States would benefit regionally by downplaying the U.S.-led Global War on Terror and placing more emphasis on its normal relations, especially since there is a perception that America is only interested in fighting Islamic militants, especially in Kenya and Sudan.

Engage rogue oil states like Sudan using track-two diplomacy to manage conflict. Track-two diplomacy has had moderate success in managing conflict in the Near East. If the United States could engage Sudan using track-two diplomacy, political tensions could be reduced, and trust could be built. However, the success or failure of track-two diplomacy depends on whether the participants are open-minded about change and have access to official policymaking circles.

Involve China in joint naval efforts to patrol coastal Africa in addition to the Horn of Africa to contribute to the stability of sea lines of communications. Joint naval cooperation to patrol coastal Africa along the SLOCs would reduce tension and build trust between the United States and China. However, China operates independently rather than jointly around the Horn of Africa for the prosecution of international pirates.

Encourage international partnerships between national oil companies and independent oil companies that demonstrate the win-win potential of international cooperation in oil investment in Africa. Encouraging international partnerships between national oil companies and international oil companies in Africa not only demonstrates the win-win potential of international cooperation but also helps to solidify the rules of the game with respect to oil energy investments.

Create patterns of cooperation with China on regional security issues such as energy security so China's growing role is in the direction of cooperation rather than conflict. China's role in Africa is rising in conjunction with increased energy demand and consumption, which may or may not have adverse consequences for bilateral relations. It would be beneficial to the United States to find novel ways, both formally and informally, to encourage China to participate in confidence-building measures (CBMs) to reduce tensions and build trust, especially since Africa produces approximately 30 percent of China's oil.

Ensure regular, senior-level contact with Chinese and African leaders to anticipate and preempt potential areas of conflict and disagreement. Regular, senior-level contact with Chinese and African leaders, especially military leaders in the form of military-to-military relations, serve as important mechanisms for building trust and reducing tensions with respect to energy security matters, especially since many oil exploration and production activities are in politically unstable regions like Nigeria and the Sudan.

WESTERN EUROPE (NORWAY AND THE UNITED KINGDOM)

Western Europe is the least energy-rich region of the world with approximately 1 percent of the world's proven oil reserves. Nevertheless, Western Europe is important to China's strategic interests. Given the lack of trust between the United States and China with respect to energy security, it would behoove the United States to undertake key multilateral initiatives in region to build trust between the United States and China. Such initiatives that the United States should consider include:

- Ensure regular, senior-level contact with Chinese and European leaders to anticipate and preempt potential areas of conflict and disagreement; and
- Promote and organize multilateral cooperative dialogues and exchanges on oil-saving technologies (particularly transportation technology).

Each of initiatives listed above is self-explanatory, but it is helpful to explain them further in terms of their benefits to United States energy security.

Ensure regular, senior-level contact with Chinese and European leaders to anticipate and preempt potential areas of conflict and disagreement. Regular, senior-level contact with Chinese and European leaders, especially military leaders in the form of military-to-military relations, serve as important mechanisms for building trust and reducing tensions with respect to energy security matters.

Promote and organize multilateral cooperative dialogues and exchanges on oil-saving technologies (particularly transportation technology). U.S.-led promotional and organizational efforts in multilateral cooperative dialogues and exchanges on oil-saving technologies in addition to unilateral domestic efforts (see above) would help to create a perception that the United States is serious about alternative fuel technology not only in the United States but also in other developed regions on the world like Western Europe.

WESTERN HEMISPHERE

The Western Hemisphere is the most energy-rich region in the world after the Near East with approximately 25 percent of the world's proven oil reserves. Further, it produces 50 percent of America's oil. Although the Western Hemisphere produces less than 7 percent of China's oil, it is important to China's energy security and strategic interests. Given the lack of trust between the United States and China with respect to energy security, China's growing influence in the region with major oil producers like Venezuela, and the region's critical importance to U.S. energy security, it would behoove the United States to undertake key unilateral, bilateral, and multilateral initiatives in region not only to build trust between the United States and China but to minimize the potential for accidental war since the region is critical to U.S. energy security. Such unilateral, bilateral, and multilateral initiatives that the United States should consider include:

- Create a perception that the United States is engaged in Latin American affairs and not just the Global War on Terror;
- Engage rogue oil states like Cuba using track-two diplomacy to manage conflict;
- Encourage international partnerships between national oil companies and independent oil companies that demonstrate the win-win potential of international cooperation in oil investment in Latin America;
- Create patterns of cooperation with China with respect to energy trading so China's growing role is in the direction of cooperation rather than conflict; and
- Ensure regular, senior-level contact with Chinese and Latin American leaders to anticipate and preempt potential areas of conflict and disagreement.

Each of the initiatives listed above is self-explanatory, but it is helpful to explain them further in terms of their benefits to United States energy security.

> *Create a perception that the United States is engaged in Latin American affairs and not just the Global War on Terror.* The Global War on Terror (GWOT) dominated the Bush Administration's foreign policy agenda in the wake of the events of September 11, 2001. While combating international terrorism made America feel safe again domestically, it hurt America's international relations. The United States would benefit regionally by downplaying the U.S.-led Global War on Terror and placing more emphasis on its normal relations, especially since there is a perception that America is only interested in fighting Islamic militants.
>
> *Engage rogue oil states like Cuba using track-two diplomacy to manage conflict.* Track-two diplomacy has had moderate success in managing conflict in the Near East and South Asia. If the United States could engage Cuba using track-two diplomacy, political tensions could be reduced, and trust could be built, especially since Cuba is potentially oil rich, and U.S. oil companies are legally prohibited from drilling within 90 miles of the U.S. coastline. However, the success or failure of track-two diplomacy depends on whether the participants are open-minded about change and have access to official policymaking circles.
>
> *Encourage international partnerships between national oil companies and independent oil companies that demonstrate the win-win potential of international cooperation in oil investment in Latin America.* Encouraging international partnerships between national oil companies and international oil companies in Latin America not only demonstrates the win-win potential of international cooperation but also helps to solidify the rules of the game with respect to oil energy investments.
>
> *Create patterns of cooperation with China with respect to energy trading so China's growing role is in the direction of cooperation rather than conflict.*

China's role in Latin America is rising in conjunction with increased energy demand and consumption, which may or may not have adverse consequences for bilateral relations. It would be beneficial to the United States to find novel ways, both formally and informally, to encourage China to participate in confidence-building measures (CBMs) to reduce tensions and build trust, especially since the region is vital to U.S. energy security.

Ensure regular, senior-level contact with Chinese and Latin American leaders to anticipate and preempt potential areas of conflict and disagreement. Regular, senior-level contacts with Chinese and African leaders serve as important mechanisms for building trust and reducing tensions with respect to energy security matters, especially since China is increasing its political and economic presence in Latin America.

U.S. POLICY RESPONSES FOR THE SURPLUS SCENARIOS

For the most part, the surplus scenarios will have positive political, economic, and military implications for the United States if they should come to fruition. The United States should act accordingly by implementing bilateral and multilateral policies that are consistent with its national security interests. However, any U.S. policy initiative should be judicious in nature and cautious in approach and should not diminish or damage Sino-American diplomatic ties, economic and trade relations, or global trade. Otherwise, the consequences will be grave for global political and economic stability as well as Sino-American relations.

In conjunction with its public policy, the United States should adopt a sophisticated strategy that utilizes the full array of diplomatic, information, economic. and military tools available in a reasonably coordinated fashion to influence China's energy outcomes. The goals of such a strategy should be to fundamentally build the mutual trust necessary for effective cooperation and to help China develop specific technical and policy capabilities that will better enable it to achieve desirable energy outcomes.[44]

RECOMMENDATIONS FOR BILATERAL INITIATIVES IN ENERGY COOPERATION

The United States and China have already undertaken a substantial number of bilateral cooperative efforts in the energy domain under the aegis of the Fossil Energy Protocol.[45] However, a high level of engagement is critical because no sector is more intimately tied to the fate of the global economy than energy, particularly oil energy.

To this extent, the United States should:

- *Phase One:* Convene bilateral dialogues on energy cooperation to build trust.
  Bilateral dialogues between the United States and China would be an important first step to build trust in energy cooperation. Dialogues would be non-statutory in nature, would be conducted formally or informally, and would be based on shared concerns on energy issues affecting both countries such as demand, consumption, and security.

- *Phase Two*: Promote cooperative dialogues on oil-saving technologies (particularly transportation technology)
  In addition to promoting bilateral dialogues, the United States can also promote specific dialogues on advanced oil-saving technologies as the next step in energy cooperation between the United States and China. Dialogues should promote the continued development and potential use of plug-in hybrid engines in addition to hybrid engines, the sustainable production of biofuels, and fuel efficiency improvements with the shared goal of reducing demand growth of fossil fuels.

- *Phase Three*: Encourage greater military-to-military ties with China, especially regarding the observation and participation in peacetime military training, exercises, and operations related to the area of energy security.
  Military cooperation, especially naval cooperation, in the area of energy security would help to build trust between China and the United States, since the potential for an accidental war is great in the East China Sea and South China Sea.

RECOMMENDATIONS FOR MULTILATERAL INITIATIVES IN ENERGY COOPERATION

Multilateral initiatives offer a very important basis for building trust in the energy world in light of the many problems in Sino-American bilateral relations. Therefore, it behooves the United States to consider and promote the following initiatives because each has the potential to produce positive outcomes in energy cooperation and to reduce Chinese mistrust of long-term U.S. intentions in the energy world. To this extent, the United States should:

- *Phase One*: Convene multilateral dialogues on energy cooperation to build trust.
  Multilateral dialogues conducted in international forums like the United Nations and the World Trade Organization would be an important initial step to

building trust in energy cooperation between the United States and China. The advantage to conducting multilateral dialogues in international forums like the United Nations and the World Trade Organization as opposed to the G-8 and the International Energy Agency is that China is already a member of the two former organizations. Membership to the latter two organizations requires China to become an advanced industrialized economy in the case of the G-8 or a member of the Organization for Economic Cooperation and Development (OECD) in the case of the International Energy Agency. To qualify for membership in the OECD, China must have a per capita GDP of at least $3,000 and meet certain standards of human rights. Therefore, "any effort to relax existing conditions to accommodate Chinese membership would prove to be too divisive among current IEA members."[46]

- *Phase Two*: Encourage international partnerships between national oil companies and independent oil companies that demonstrate the win-win potential of international cooperation in oil investment (*competitive surplus scenario*)
  Encouraging international partnerships between national oil companies and international oil companies for projects in mainland China and China's territorial waters not only demonstrates the win-win potential of international cooperation but also helps to solidify the rules of the game with respect to oil energy investments.

- *Phase Three*: Broaden diplomatic dialogue inside regions where there are significant oil energy investments to identify overlapping interests, such as stable investment conditions, and possible areas for cooperation, including foreign assistance (*competitive curplus scenario*).
  Broadening diplomatic dialogues in oil-rich regions of East Asia and Southeast Asia to identify overlapping interests, such as stable investment conditions, and possible areas for cooperation, including foreign assistance with respect to oil energy investments, can bring about conflict resolution as well as building trust and cooperation between the United States, China, and oil energy producers.

RECOMMENDATIONS FOR REGIONAL INITIATIVES IN ENERGY COOPERATION

In surplus scenarios, regional initiatives are proposed for the Asia-Pacific region. The initiatives will vary in size and scope in light of overall U.S. and Chinese strategic engagement and available oil energy reserves in each region.

THE ASIA-PACIFIC

The Asia-Pacific is one of the least energy-rich regions of the world, with approximately 7 percent of the world's proven oil reserves. Nevertheless, Asia is vital to China's strategic interests. Given the potential for a kinetic military exchange in the East China Sea and South China Sea, it would behoove the United States to undertake key unilateral, bilateral, and multilateral initiatives in the region not only to build trust between the United States and China but to minimize the potential for accidental war. Such initiatives that the United States should consider include:

- Create patterns of cooperation with China on regional security issues such as energy security so China's growing role is in the direction of cooperation rather than conflict.
  Chinese as well as U.S. oil energy demand and consumption will increase over the next 20 years. China's role on the international stage is expected to rise in conjunction with increased energy demand and consumption, which may or may not have adverse consequences for bilateral relations. It would be beneficial to the United States to find novel ways, both formally and informally, to encourage China to participate in confidence-building measures (CBMs) to reduce tensions and the specter of an accidental war in the region.

- Empower regional institutions, such as the ASEAN Regional Forum (ARF), through forms of multilateral cooperation to address energy security challenges. The ASEAN Regional Forum (ASF) has been regarded as a better model for security cooperation in non-Western regions than the European experience of highly institutionalized regional cooperation because of its emphasis on personal contacts, informality, and consensus-building.[47] Consisting of 27 participants, the ARF objectives are to foster dialogue and consultation and promote confidence-building and preventive diplomacy in the region.

Notes

1. Downs, Erica Strecker, *China's Energy Security* (UMI Dissertation), 2004, 243.

2. Saunders, Philip C., *China's Global Activism: Strategy, Drivers, and Tools* (Washington, D.C.: Institute for National Strategic Studies, National Defense University), 2006, 28.

3. Military Advisory Board, *Powering America's Defense: Energy and the Risks to National Security* (Alexandria, VA: Center for Naval Analysis, May 2009), 1.

4. *USA Today*, "Chinese Economy Slows To Still Sizzling 11.5% Growth," September 25, 2007.

5. Lieberthal, Kenneth and Mikkal Herberg, *China's Search for Energy Security: Implications for U.S. Policy* (Seattle: National Bureau of Asian Research), Volume 17, Number 1, April 2006, 28.

6. Lieberthal, Kenneth and Mikkal Herberg, *China's Search for Energy Security: Implications for U.S. Policy*, 28.

7. Lieberthal, Kenneth and Mikkal Herberg, *China's Search for Energy Security: Implications for U.S. Policy*, 29.

8. Lieberthal, Kenneth and Mikkal Herberg, *China's Search for Energy Security: Implications for U.S. Policy* (Seattle: National Bureau of Asian Research), Volume 17, Number 1, April 2006, 28.

9. Lieberthal, Kenneth and Mikkal Herberg, *China's Search for Energy Security: Implications for U.S. Policy* (Seattle: National Bureau of Asian Research), Volume 17, Number 1, April 2006, 29.

10. Lieberthal, Kenneth and Mikkal Herberg, *China's Search for Energy Security: Implications for U.S. Policy* (Seattle: National Bureau of Asian Research), Volume 17, Number 1, April 2006, 29–30.

11. Lieberthal, Kenneth and Mikkal Herberg, *China's Search for Energy Security: Implications for U.S. Policy* (Seattle: National Bureau of Asian Research), Volume 17, Number 1, April 2006, 30.

12. Lieberthal, Kenneth and Mikkal Herberg, *China's Search for Energy Security: Implications for U.S. Policy* (Seattle: National Bureau of Asian Research), Volume 17, Number 1, April 2006, 30.

13. Lieberthal, Kenneth and Mikkal Herberg, *China's Search for Energy Security: Implications for U.S. Policy* (Seattle: National Bureau of Asian Research), Volume 17, Number 1, April 2006, 30.

14. Lieberthal, Kenneth and Mikkal Herberg, *China's Search for Energy Security: Implications for U.S. Policy* (Seattle: National Bureau of Asian Research), Volume 17, Number 1, April 2006, 30.

15. Evans, Peter C. and Erica S. Downs, *Untangling China's Quest for Oil through State-backed Financial Deals* (The Brookings Institution), Policy Brief 154, May 2006, 6.

16. Evans, Peter C. and Erica S. Downs, *Untangling China's Quest for Oil through State-backed Financial Deals*, 6.

17. Organization for Economic Co-Operation and Development, *The Export Credits Arrangement*, 1978–2008.

18. Organization for Economic Co-Operation and Development, *The Export Credits Arrangement*, 1978–2008.

19. Evans, Peter C. and Erica S. Downs, *Untangling China's Quest for Oil through State-backed Financial Deals*, 7.

20. Lieberthal, Kenneth and Mikkal Herberg, *China's Search for Energy Security: Implications for U.S. Policy* (Seattle: National Bureau of Asian Research), Volume 17, Number 1, April 2006, 33.

21. Lieberthal, Kenneth and Mikkal Herberg, *China's Search for Energy Security: Implications for U.S. Policy* (Seattle: National Bureau of Asian Research), Volume 17, Number 1, April 2006, 32.

22. Lieberthal, Kenneth and Mikkal Herberg, *China's Search for Energy Security: Implications for U.S. Policy* (Seattle: National Bureau of Asian Research), Volume 17, Number 1, April 2006, 32.

23. Lieberthal, Kenneth and Mikkal Herberg, *China's Search for Energy Security: Implications for U.S. Policy* (Seattle: National Bureau of Asian Research), Volume 17, Number 1, April 2006, 34.

24. Crane, Keith, Andreas Goldthau, Michael Toman, Thomas Light, Stuart E. Johnson, Alireza Nader, Angel Rabasa, and Harun Dogo, *Imported Oil and U.S. National Security* (Santa Monica, California: RAND), 2009, xiii.

25. Crane, Keith, Andreas Goldthau, Michael Toman, Thomas Light, Stuart E. Johnson, Alireza Nader, Angel Rabasa, and Harun Dogo, *Imported Oil and U.S. National Security*, 82.

26. Crane, Keith, Andreas Goldthau, Michael Toman, Thomas Light, Stuart E. Johnson, Alireza Nader, Angel Rabasa, and Harun Dogo, *Imported Oil and U.S. National Security*, 78.

27. Crane, Keith, Andreas Goldthau, Michael Toman, Thomas Light, Stuart E. Johnson, Alireza Nader, Angel Rabasa, and Harun Dogo, *Imported Oil and U.S. National Security*, 79.

28. Crane, Keith, Andreas Goldthau, Michael Toman, Thomas Light, Stuart E. Johnson, Alireza Nader, Angel Rabasa, and Harun Dogo, *Imported Oil and U.S. National Security*, 79.

29. Crane, Keith, Andreas Goldthau, Michael Toman, Thomas Light, Stuart E. Johnson, Alireza Nader, Angel Rabasa, and Harun Dogo, *Imported Oil and U.S. National Security*, 80.

30. National Petroleum Council, *Facing The Hard Truths About Energy: A Comprehensive View To 2030 Of Global Oil And Natural Gas (Draft Report)* (Washington, D.C.), July 18, 2007, 120.

31. National Petroleum Council, *Facing The Hard Truths About Energy: A Comprehensive View To 2030 Of Global Oil And Natural Gas (Draft Report)* (Washington, D.C.), July 18, 2007, 120.

32. National Petroleum Council, *Facing The Hard Truths About Energy: A Comprehensive View To 2030 Of Global Oil And Natural Gas (Draft Report)* (Washington, D.C.), July 18, 2007, 120.

33. National Petroleum Council, *Facing The Hard Truths About Energy: A Comprehensive View To 2030 Of Global Oil And Natural Gas (Draft Report)* (Washington, D.C.), July 18, 2007, 121.

34. Sandalow, David, *Ending Oil Dependence* (Washington, D.C.: The Brookings Institution), January 22, 2007, 14.

35. Sandalow, David, *Ending Oil Dependence*, 14.
36. Sandalow, David, *Ending Oil Dependence*, 15.
37. Sandalow, David, *Ending Oil Dependence*, 15.
38. Sandalow, David, *Ending Oil Dependence*, 15.
39. Sandalow, David, *Ending Oil Dependence*, 15.
40. Sandalow, David, *Ending Oil Dependence*, 19.
41. Kaye, Dalia Dassa, *Talking to the Enemy: Track-Two Diplomacy in the Middle East and South Asia* (Santa Monica, California: RAND), MG-592-NSRD, 2007, 93.
42. Kaye, Dalia Dassa, *Talking to the Enemy: Track-Two Diplomacy in the Middle East and South Asia*, 93.
43. Kaye, Dalia Dassa, *Talking to the Enemy: Track-Two Diplomacy in the Middle East and South Asia*, 13.
44. Lieberthal, Kenneth and Mikkal Herberg, *China's Search for Energy Security: Implications for U.S. Policy* (Seattle: National Bureau of Asian Research), Volume 17, Number 1, April 2006, 28.
45. http://fossil.energy.gov/international/International_Partners/China.html.
46. Lieberthal, Kenneth and Mikkal Herberg, *China's Search for Energy Security: Implications for U.S. Policy* (Seattle: National Bureau of Asian Research), Volume 17, Number 1, April 2006, 33.
47. Kaye, Dalia Dassa, *Talking to the Enemy: Track-Two Diplomacy in the Middle East and South Asia*, 13.

Appendix A

World's Proved Reserves of Oil by Region (Most Recent Estimates)

Table A.1 World's Proved Reserves of Oil by Region

Region	*Country*	*Oil and Gas Journal Oil (Billion Barrels)*	*Oil (Percentage) of World Total*
Asia & Oceania	Australia	1.500	0.11
Asia & Oceania	Brunei	1.100	0.08
Asia & Oceania	Burma (Myanmar)	0.050	0.00
Asia & Oceania	China	16.000	1.19
Asia & Oceania	Indonesia	3.990	0.30
Asia & Oceania	Korea, South	0	0.00
Asia & Oceania	Malaysia	4.000	0.30
Asia & Oceania	Mongolia	0	0.00
Asia & Oceania	Philippines	0.139	0.01
Asia & Oceania	Thailand	0.441	0.03
Asia & Oceania	Vietnam	0.600	0.04
	Total	27.820	2.07
Eurasia	Azerbaijan	7.000	0.52
Eurasia	Kazakhstan	30.000	2.24
Eurasia	Russia	60.000	4.47
Eurasia	Turkmenistan	0.600	0.04
Eurasia	Uzbekistan	0.594	0.04
	Total	98.194	7.32

(*Continued*)

Table A.1 **World's Proved Reserves of Oil by Region (Continued)**

Region	*Country*	*Oil and Gas Journal Oil (Billion Barrels)*	*Oil (Percentage) of World Total*
Middle East & North Africa	Algeria	12.200	0.91
Middle East & North Africa	Egypt	3.700	0.28
Middle East & North Africa	Iran	136.150	10.14
Middle East & North Africa	Iraq	115.000	8.57
Middle East & North Africa	Kuwait	104.000	7.75
Middle East & North Africa	Libya	43.660	3.25
Middle East & North Africa	Morocco	0.001	0.00
Middle East & North Africa	Oman	5.500	0.41
Middle East & North Africa	Qatar	15.210	1.13
Middle East & North Africa	Saudi Arabia	266.710	19.87
Middle East & North Africa	Syria	2.500	0.19
Middle East & North Africa	Tunisia	0.425	0.03
Middle East & North Africa	United Arab Emirates	97.800	7.29
Middle East & North Africa	Yemen	3.000	0.22
	Total	805.856	60.04
Sub Saharan Africa	Angola	9.040	0.67
Sub Saharan Africa	Chad	1.500	0.11
Sub Saharan Africa	Congo (Brazzaville)	1.600	0.12
Sub Saharan Africa	Cote d'Ivoire (Ivory Coast)	0.100	0.01
Sub Saharan Africa	Equatorial Guinea	1.100	0.08
Sub Saharan Africa	Gabon	2.000	0.15
Sub Saharan Africa	Guinea	0	0.00
Sub Saharan Africa	Kenya	0	0.00
Sub Saharan Africa	Madagascar	0	0.00
Sub Saharan Africa	Mali	0	0.00
Sub Saharan Africa	Mauritania	0.100	0.01
Sub Saharan Africa	Niger	0	0.00
Sub Saharan Africa	Nigeria	36.220	2.70
Sub Saharan Africa	South Africa	0.015	0.00
Sub Saharan Africa	Sudan	5.000	0.37
	Total	56.675	4.22

(*Continued*)

Table A.1 World's Proved Reserves of Oil by Region (Continued)

Region	*Country*	*Oil and Gas Journal Oil (Billion Barrels)*	*Oil (Percentage) of World Total*
Western Europe	Norway	6.680	0.50
Western Europe	United Kingdom	3.410	0.25
	Total	10.090	0.75
Western Hemisphere	Canada	178.092	13.27
	Mexico	10.501	0.78
	Argentina	2.616	0.19
	Bolivia	0.465	0.03
	Brazil	12.624	0.94
	Colombia	1.355	0.10
	Cuba	0.124	0.01
	Ecuador	4.660	0.35
	Guatemala	0.083	0.01
	Peru	0.416	0.03
	Venezuela	99.377	7.40
	Total	310.313	23.11
All Regions	Sub Total	1,130.856	97.52
	World Total	1,342.207	100.00

Source: Energy Information Agency, U.S. Department of Energy, from Oil & Gas Journal as of January 1, 2009.

Appendix B

Research Design

Scenario analysis provides the best way to establish quantitative causality between China's alternative futures, or scenarios (and associated events), and the possible strategic implications for the United States. Scenario analysis is a process of analyzing possible future events by considering alternative possible outcomes.

> Through the development and use of scenarios, we challenge our assumptions and can be better prepared for the unexpected when it inevitably occurs. [1] "Scenario analysis focuses on the areas of greatest uncertainty for a country or an operation, systematically develops several plausible alternative future environments in which the operation might be implemented, and determines how they would affect its success . . . The analysis is designed to allow improved decision-making by allowing more complete consideration of outcomes and their implications. . . . In politics or geo-politics, scenario analysis involves modeling the possible alternative paths of a social or political environment and possibly diplomatic and war risks."[2]

In addition, the PRINCE forecasting method that was developed by William D. Coplin and Michael K. O'Leary is used in this book to determine whether China and the United States will peacefully share the world's oil reserves or whether competition for oil and gas reserves will lead to conflict.

According to William D. Coplin and Michael K. O'Leary, the PRINCE model is useful in a wide variety of political situations and is one of the few approaches to political forecasting that is both systematic and relevant to decision-makers. The model provides a means for the rigorous analysis of information, producing calculations that result in forecasts of the probability of specific outcomes. At the same time, it uses the invaluable and unique expertise of specialists, both to supply information about the model and to adjust the calculations produced by the model according to qualitative, subjective knowledge about a given country.[3]

In the PRINCE Model, the actors (or stakeholders) are identified by their power, position, and salience, which are used to calculate the probabilities for each scenario's associated outcome, determining which energy future the Chinese are most likely to adopt. Stakeholders are identified based on the existing body of literature, their domestic and international political relationships and interactions with other stakeholders, and the value of their available resources—which equates to their relative power—and their position on the issue of Chinese oil futures and their salience for the issue of oil competition given all other important issues that each stakeholder must deal with, will be estimated. These factors are assigned a numerical value based on the perspectives of survey participants who have knowledge of Chinese oil energy politics. Surveys were sent to more than 400 experts in the United States, Asia. and Western Europe. Survey participants were individuals from academia, government, and the private sector who have specialized knowledge of any of the following: oil politics and/or economics, Chinese oil energy politics, Chinese politics and/or economics, Chinese maritime security issues, the comparative politics of Asia, Africa, Europe (including Russia), Latin America, North America, and the Near East as it relates to oil or national security issues.

Event identification and assignation to a particular Chinese oil energy scenario is determined by actual and potential world events and whether the event directly involves the supply of oil from its country or region of origin or indirectly involves the supply of oil through a third country or through a national security event, such as Sino-Indian military relations or U.S.-Indian military relations.

In his seminal work *Expert Political Judgment: How good is it? How can we know?*, Philip Tetlock (2005) found that experts in foreign affairs make poor predictions about future events and are no better forecasters than informed non-experts in making predictions. Tetlock also observed that experts tend to overestimate the probability of likely events that will occur as well as overestimate the probability of unlikely events that will not occur, resulting in less accurate predictions of future events. Similarly, in "How Statistical Models Can Help Intelligence Analysts," Steven Rieber (2004) observed that political experts tend to misjudge not only which factors are relevant but how they are relevant. He observed that experts will often fail to predicate their prediction estimates on a large sample of relevant data, relying instead on small numbers of memorable or anecdotal illustrations. Reiber suggested that formal models can usefully supplement expert judgment because they test the validity of factors/variables and their assigning weights; base their predictions on all known relevant data such as statistical samples; minimize both random and systematic error; establish useful baselines; deepen experts' understanding of key entities, classes, concepts and relationships; and reveal underlying causality in phenomena.

Formal models are also easy to use but require data. Experts can and do provide data. When expert data is used and properly calibrated, more accurate predictions of future outcomes will result. Otherwise, in the absence of using expert data and a formal model for forecasting futures such as the PRINCE Model, experts will tend to overestimate the probability of likely events that will occur as well as overestimate the probability of unlikely events that will not occur, resulting in less accurate predictions of future events.

THE MECHANICS OF THE PRINCE MODEL

PRINCE is an acronym for four steps: "**Pr**obe, **In**teract, **C**alculate, and **E**xecute." Each step is integral to the PRINCE model of risk analysis forecasting. The following discussion is indebted to William Coplin, Pat McGowan and Michael O'Leary's *Everyman's Prince: A Guide to Understanding Your Political Problems* (1972).

Step 1: Probe

In the probe step policy experts define the issue and identify the actors. An issue is a proposed decision or action that is likely to generate controversy. The policy experts surveyed should include a cross-range of theorists and practitioners who have first-hand knowledge of the issue at hand. An actor is any individual, group, or organization that ought to be considered in making the decision or in carrying it out after it has been made. This information is necessary to analyze each actor's influence on the issue.

Step 2: Interact

Once the actors and issues have been identified by policy experts the next step is to analyze each actor's influence on the issue. Influence is the combination of position, salience, and power. These three variables are necessary to calculate the actors' influence.

- *Position* is whether one favors or opposes a particular solution to an issue. Power is measured on a scale of -5 to +5 (-5, -4, -3, -2, -1, 0, 1, 2, 3, 4, and 5). The higher the number, the greater an actor favors (+) or opposes (-) a particular solution to an issue.
- *Salience* is the strength of the support or opposition. Salience is measured on a scale of 1 to 5 (1, 2, 3, 4, and 5). The higher the number, the greater the strength of the support or opposition of the decision-making actor.
- *Power* is the ability to make one's preference on an issue take place. Power is measured on a scale of 1 to 5 (1, 2, 3, 4, and 5). The higher the number, the greater actor's ability to actualize his/her preference.

Policy experts estimated each actor's position, salience and power relative to the other actors making the decision. Each actor's relative influence (the combination of position, salience and power) is then recorded on a matrix (see Step 3 below). Ideally, a sample size of at least 30 policy experts is recommended to participate in a PRINCE survey.

Step 3: Compute

Once information has been gathered on each actor's position, salience, and power, the next step is to build a series of tables summarizing the findings. This is a two-step process of determining each actor's weight and calculating the probability of the decision's being adopted.

Position, Salience, and Power Matrix: The actors are listed in the first column and the policy issue across the top of the matrix and the numerical values entered for each factor. Actors with no position (either no position yet or a firm "I don't care") on an issue are assigned a zero for that issue. The position variable may be positive (+) or negative (-), while the other variables (power and salience) are always positive. The sum of the actor's positions on each issue elicits how much overall support there is for each issue. A strong opposition countered by a strong proponent has the same effect as two undecided positions, so the value of the sum may be less informative than how it was constituted.

Once the estimates are made for each actor, the next step is to calculate the weights each actor contributes to the decision. This is done by multiplying position times salience times power for each actor. After each actor's weight (position times salience times power) is calculated, the sum of the actors' weights is determined by adding them.

The next step is to estimate the probability of the decision to be adopted. First, the scores of all the actors supporting the decision are added, and this column total is A. The scores of all the actors opposing the issue are added, and this column total is B. Minus signs are eliminated to create the absolute value. The non-zero scores of all the actors that have a neutral position are multiplied and then added together in Column C. Columns A, B, and C are added across for each actor and recorded in Column D, giving the total of all the power weights in the particular issue.

Next, Column A values (the strongest supporters and opponents) are added to one-half of the value of Column C, the neutral actors' scores (recorded in Column E). This group of actors constitutes the "swing voters." that is, the neutral actors who are equally likely, in the future, to be either supporters or opponents of the issue. The best way to represent this 50–50 situation in the absence of other information is to include just half the neutral actors' scores with the positive weights. This total is the sum of the best estimate of the likely weights to be exerted in support of the issue.

The total of weights supporting the issue (E) is divided by the total of weights in the system (D). The resulting fraction is the proportion of the positive weights in relation to the total weights and ranges from 0.0 (no chance of occurrence) to 1.00 (certainty of occurrence). It can be interpreted as the probability that the issue will be implemented. If the probability is between 0 percent and 39 percent, the likelihood of occurrence is low; if the probability is between 40 percent and 59 percent, the likelihood of occurrence is uncertain; and if the probability is between 60 percent and 100 percent, the likelihood of occurrence is high. See the equation below.

$$\text{PRINCE Probability Equation} = \frac{\Sigma W_s}{\Sigma W_t} = \frac{\text{Sum of Total Positive Actor Scores}}{\text{Sum of Absolute Value of All Actor Scores}}$$

Step 4: Execute

The last step is to formulate and execute a strategy. On the assumption that the outcome is not a "sure thing." there are only four tools to increase the likelihood of the desired outcome: 1. Change the position of one or more actors on the issue; 2. Increase the salience of the issue for those actors who side with you or decrease it for those who oppose you; 3. Build the power of those actors who side with you or weaken the power of those who oppose you; 4. Make friends. and win over your enemies.

Notes

1. Attributed to James Burkhard of IHS Cambridge Energy Research Associates (CERA).

2. Mack, Jonathan N., "Scenario Analysis: A Tool for Task Managers," in *Social Analysis: Selected Tools and Techniques* (Washington, D.C.: World Bank), 2001, 62.

3. Coplin, William D. and Michael K. O'Leary, *Introduction to Political Risk Analysis (Learning Packages in the Policy Sciences PS-24)(* New York: Policy Studies Associates), December 1983, 18.

Selected Bibliography

Alterman, Jon. B. and John W. Garver. The Vital Triangle: China, The United States and the Middle East. Washington, D.C.: CSIS Press, 2008.

Andrews-Speed, Philip, *Energy Policy And Regulation In The People's Republic Of China*. The Hague, The Netherlands: Kluwer Law International, 2004.

Andrews-Speed, Philip, Xuanli Liao and Roland Dannreuther, *The Strategic Implications Of China's Energy Needs*. New York, Oxford University Press, 2002.

Asian Pacific Research Centre, *Energy In China: Transportation, Electrical Power And Fuel Markets*. Tokyo, Japan: Asian Pacific Research Centre, 2004.

Bajpaee, Chietigj, "Sino-US Energy Competition in Africa," October 7, 2005, www.globalpolicy.org/security/natres/oil/2005/1007competition.htm.

Berman, Ilan, "The Impact of Sino-Iranian Strategic Partnership" (Testimony prepared for the U.S.-China Economic & Security Review Commission), September 14, 2006.

Bhalla, Madhu, "Geopolitics of Economic Relations" in Swaran Singh, ed., *China-Pakistan Strategic Cooperation: Indian Perspectives*. New Delhi, India: Manohar Publishers, 2007, pp. 251–274.

Bliss, Catherine E., "China's Projection of Soft Power in the Americas", in Carola McGiffert, ed., *Chinese Soft Power and Its Implications for the United States*. Washington, D.C.: CSIS Press, March 2009, pp. 45–62.

Blumenthal, Dan and Joseph Lin, "Oil Obsession: Energy Appetite Fuels Beijing's Plans to Protect Vital Sea Lanes," www.armedforcesjournal.com, June 2006.

British Petroleum, *Energy Trends and Technologies for the Coming Decades: Address to the Harvard University Center for the Environment*, British Petroleum, March 8, 2006.

Cameron, Fraser and Yongnian Zheng, "Key Elements of a Strategic Partnership", in Stanley Crossick and Etienne Reuter, eds., *China-EU: A Common Future*. Singapore: World Scientific Publishing, 2007, pp. 3–14.

Campos, Indira, Michael Chege, Pat Utomi and Alex Vines, *U.S. and Chinese Engagement in Africa: Prospects for Improving U.S.-China Africa Cooperation*, Washington, D.C.: CSIS Press, July 2008.

Carman, Jessie C., "Economic and Strategic Implications of Ice-Free Arctic Seas," in Sam J. Tangredi, ed., *Globalization and Maritime Power.* Honolulu: University Press of the Pacific, 2004, pp. 171–188.

Chambers, Michael, "China's Military Rise to Great Power Status: Its Implications for the United States in Southeast Asia," in Evelyn Goh and Sheldon W. Simon, eds., *China, United States, and Southeast Asia: Contending Perspectives on Politics, Security, and Economics.* New York: Routledge, 2008, pp. 167–184.

Cole, Bernard D., *Oil for the Lamps of China-Beijing's 21st Century Search for Energy*. Washington, D.C.: Institute for National Strategic Studies, National Defense University, 2003.

Cole, Bernard D., *Sea Lanes and Pipelines: Energy Security in Asia.* Westport, Connecticut: Praeger Security International, 2008.

Congressional Research Service, Library of Congress, *China's Foreign Policy and "Soft Power" in South America, Asia, and Africa: A Study Prepared for the Committee on Foreign Relations United States Senate.* Washington, D.C.: U.S. Government Printing Office, April 2008.

Congressional Research Service, Library of Congress, *China's "Soft Power" in Southeast Asia.* Washington, D.C.: U.S. Government Printing Office, RL 34310, January 4, 2008.

Congressional Research Service, Library of Congress, *Comparing Global Influence: China's and U.S. Diplomacy, Foreign Aid, Trade, and Investment in the Developing World.* Washington, D.C.: U.S. Government Printing Office, RL 34620, August 15, 2008.

Cooney, Jim, *Chinese Oil Dependence: Opportunities and Challenges.* Carlisle, U.S. Army War College, March 18, 2005.

Cooper, Cortez A., *The PLA Navy's "New Historic Missions": Expanding Capabilities for a Re-emergent Maritime Power.* Santa Monica, RAND, CT-332, June 11, 2009.

Cooper, William H., *EU-U.S. Economic Ties: Framework, Scope, and Magnitude.* Congressional Research Service, the Library of Congress, RL30608, January 3, 2008.

Coplin, William D. and Michael K. O'Leary, *Introduction to Political Risk Analysis (Learning Packages in the Policy Sciences PS-24).* New York: Policy Studies Associates, December 1983.

Coplin, William D., Pat McGowan and Michael K. O'Leary, *Everyman's Prince: A Guide to Understanding Your Political Problems.* Pacific Grove, California: Duxbury Press, 1972.

Coulter, Donald Y., "Globalization of Maritime Commerce: The Rise of Hub Ports," in Sam J. Tangredi, ed., *Globalization and Maritime Power.* Honolulu: University Press of the Pacific, 2004, pp. 133–42.

Courtney, Hugh, Jane Kirkland and Patrick Viguerie, "Strategy Under Uncertainty," *Harvard Business Review*, November-December 1997, pp. 66–79.

Crane, Keith, Andreas Goldthau, Michael Toman, Thomas Light, Stuart E. Johnson, Alireza Nader, Angel Rabasa and Harun Dogo, *Imported Oil and U.S. National Security*. Santa Monica, California: RAND, 2009.

Davies, Martyn J., "Special Economic Zones: China's Developmental Model Comes to Africa" in Robert I. Rotberg, ed., *China into Africa: Trade, Aid and Influence*. Washington, D.C.: Brookings Institution Press, 2008, pp. 137–54.

Downs, Erica Strecker, *China's Energy Security*, UMI Dissertation, 2004.

Downs, Erica Strecker, *China's Quest for Energy Security*. Santa Monica, California: RAND, MR-1244-AF, 2000.

Downs, Erica Strecker, "Foreign Policy Studies Energy Security Series: China," *The Brookings Institution*, December 2006.

Dreyer, June Teufel, *Prepared Statement before the Subcommittee on the Western Hemisphere*, April 6, 2005.

Dumbaugh, Kerry and Mark P. Sullivan, *China's Growing Interest in Latin America*. Congressional Research Service, the Library of Congress, RS22119, April 20, 2005.

Ebel, Robert E., *China's Energy Future: The Middle Kingdom Seeks Its Place in the Sun*. Washington, D.C.: CSIS Press, 2005.

Eisenman, Joshua, "China's Post-Cold War Strategy in Africa: Examining Beijing's Methods and Objectives," in Joshua Eisenman, Eric Heginbotham, and Derek Mitchell, eds., *China and the Developing World: Beijing's Strategy for the Twenty-First Century*. Armonk, New York: M.E. Sharpe, 2007, pp. 29–59.

Ellis, Evan, "China's Maturing Relationship with Latin America," *China Brief*, Jamestown Foundation, Volume IX, Issue 6, March 19, 2009, pp. 4–7.

Evans, Peter C. and Erica S. Downs, *Untangling China's Quest for Oil through State-backed Financial Deals*, The Brookings Institution, Policy Brief 154, May 2006.

Flanagan, Stephen J., "China's Military Diplomacy and Engagement in Latin America", in Carola McGiffert, ed., *Chinese Soft Power and Its Implications for the United States*. Washington, D.C.: CSIS Press, March 2009, pp. 54–56.

Garver, John W., *China's South Asian Interests and Policies*, Prepared for the panel on "China's Approaches to South Asia and the Former Soviet States," U.S.-China Economic and Security Review Commission, July 22, 2005.

Gelb, Bernard A, *Unocal Corporation's Oil and Gas*. Congressional Research Service, the Library of Congress, RS22182, July 26, 2005.

Gill, Bates, Chin-hao Huang and J. Stephen Morrison, *China's Expanding Role in Africa: Implications for the United States*. Washington, D.C.: CSIS Press, January 2007.

Gill, Bates and Melissa Murphy, *China-Europe Relations: Implications and Policy Reponses for the United States*. Washington, D.C.: CSIS Press, May 2008.

Gill, Bates and Matthew Oresman, *China's New Journey to the West: China's Emergence in Central Asia and Implications for U.S. Interests*. Washington, D.C.: CSIS Press, August 2003.

Goh, Evelyn and Sheldon W. Simon, "Introduction," in Evelyn Goh and Sheldon W. Simon, eds., *China, United States, and Southeast Asia: Contending Perspectives on Politics, Security, and Economics*. New York: Routledge, 2008, pp. 1–14.

Gupta, Sonika, "Role of Political Culture" in Swaran Singh, ed., *China-Pakistan Strategic Cooperation: Indian Perspectives*. New Delhi, India: Manohar Publishers, 2007, pp. 119–36.

Herberg, Mikkal, "China's Search for Energy Security: the Implications for the Southeast Asia," in Evelyn Goh and Sheldon W. Simon, eds., *China, United States, and Southeast Asia: Contending Perspectives on Politics, Security, and Economics*. New York: Routledge, 2008, pp. 70–88.

Ho, Joshua, *The Shifting of Maritime Power and the Implications for Maritime Security in East Asia*. Singapore: Institute of Defence and Strategic Studies, No. 68, June 2004.

Hoffman, Nathaniel, "Chinese Presence in Cuba Increasing," *McClatchy Newspapers*, June 24, 2007.

Hoge, Warren, "Security Council Is Stalled Over Iran's Nuclear Program," *New York Times*, March 22, 2006, www.nytimes.com/2006/03/22/inter-national/middleeast/22iran.html?_r=2&pagewanted=print&oref=slogin.

Horta, Loro, "In Uncle Sam's Backyard: China's Military Influence in Latin America," *Military Review*, September-October 2008, pp. 47–55.

Hurst, Cindy, *China's Oil Rush In Africa*. Washington, D.C.: Institute for the Analysis of Global Security, July 2006.

International Energy Agency, *China's Worldwide Quest for Energy Security*. Paris, France: Organization for Economic Cooperation and Development/International Energy Agency, 2000.

International Energy Agency, *Energy Statistics of Non-OECD Countries, 2002–2003*. Paris, France: Organization for Economic Cooperation and Development/International Energy Agency, 2005.

International Energy Agency, *The Impact of the Financial and Economic Crisis on Global Energy Investment*. Paris, France: Organization for Economic Cooperation and Development/International Energy Agency, May 2009.

Jiang, Julie and Jonathan Sinton, *Overseas Investments by Chinese National Oil Companies*. Paris, France: Organization for Economic Cooperation and Development/International Energy Agency, February 2011.

Jiang, Wenran, "China's Emerging Strategic Partnerships in Africa" in Robert I. Rotberg, ed., *China into Africa: Trade, Aid and Influence*. Washington, D.C.: Brookings Institution Press, 2008, pp. 50–64.

Jiang, Wenran, "China's Growing Energy Relations with the Middle East," *China Brief*, Jamestown Foundation, Volume VII, Issue 14, July 11, 2007, pp. 12–15.

Jin, Liangxiang, "Sino-Arab Relations: New Developments and Trends, *Middle East Policy*, Volume XI, No. 4, Winter 2004, pp. 113–21.

Johnston, Corey S., *Transnational Pipelines and Naval Expansion: Examining China's Oil Insecurities in the Indian Ocean (Master's Thesis)*. Monterey, CA: Naval Postgraduate School, June 2008.

Kalicki, Jan H. and David L. Goldwyn, eds., *Energy and Security: Toward a New Foreign Policy Strategy.* Washington, D.C.: Woodrow Wilson Center Press, 2005.

Kaufman, Alison A., *China's Participation in Anti-Piracy Operations off the Horn of Africa: Drivers and Implications.* Alexandria, Virginia: Center for Naval Analysis, July 2009.

Kaye, Dalia Dassa, *Talking to the Enemy: Track-Two Diplomacy in the Middle East and South Asia.* Santa Monica, California: RAND, MG-592-NSRD, 2007.

King, Jr., Neil, "China-Iran Trade Vexes U.S.," *Wall Street Journal*, July 27, 2007, pg. 4.

Klare, Michael, T., *Blood and Oil: The Dangers and Consequences of America's Growing Dependency of Imported Petroleum.* New York: Owl Books, 2004.

Klare, Michael T., "Petroleum Anxiety and the Militarization of Energy Security," in Daniel Moran and James A. Russell, eds., *Energy Security and Global Politics: the Militarization of Resource Management.* New York: Routledge, 2009, pp. 39–61.

Klare, Michael, T., *Resource Wars: The New Landscape of Global Conflict.* New York: Owl Books, 2002.

Klare, Michael, T., *Rising Powers, Shrinking Planet: The New Geopolitics of Energy.* New York: Metropolitan Books, 2008.

Kong, Bo, *An Anatomy of China's Insecurity and Its Strategies.* Seattle: Pacific Northwest Center for Global Security, PNNL-15529, December 2005.

Kraemer, Thomas D., *Addicted to Oil: Strategic Implications of American Oil Policy.* Strategic Studies Institute, U.S. Army War College, May 2006.

Kronstadt, K. Alan, *India-U.S. Relations,* Congressional Research Service, Library of Congress, RL33529, January 30, 2009.

Kronstadt, K. Alan, *Pakistan-U.S. Relations,* Congressional Research Service, Library of Congress, RL33498, November 10, 2008.

Kurlantzick, Josh, "Beijing's Safari: China's Move into Africa and Its Implications for Aid, Development, and Governance," *Policy Outlook*, No. 29, November 2006.

Lal, Rollie, *Central Asia and Its Neighbors: Security and Commerce at the Crossroads.* Santa Monica, California: RAND, MG-440, 2006.

Lal, Rollie, "China's Relations with South Asia," in Joshua Eisenman, Eric Heginbotham, and Derek Mitchell, eds., *China and the Developing World: Beijing's Strategy for the Twenty-First Century.* Armonk, New York: M.E. Sharpe, 2007, pp. 133–149.

Lal, Rollie, *Understanding China and India: Security Implications for the United States and the World.* Westport: Praeger Security International, 2006.

Lanti, Irman, "Indonesia in Triangular Relations with China and the United States," in Evelyn Goh and Sheldon W. Simon, eds., *China, United States, and Southeast Asia: Contending Perspectives on Politics, Security, and Economics.* New York: Routledge, 2008, pp. 128–42.

Lee, Henry and Dan A. Shalmon, "Searching for Oil: China's Initiatives in the Middle East," *Environment*, Volume 49, Number 5, June 2007, pp. 8–21.

Lee, Henry and Dan Shalmon, "Searching for Oil: China's Oil Strategies in Africa" in Robert I. Rotberg, ed., *China into Africa: Trade, Aid and Influence.* Washington, D.C.: Brookings Institution Press, 2008, pp. 109–36.

Leverett, Flynt, "Resource Mercantilism and the Militarization of Resource Management: Rising Asia and the Future of American Primacy," in Daniel Moran and James A. Russell, eds., *Energy Security and Global Politics: the Militarization of Resource Management.* New York: Routledge, 2009, pp. 211–42.

Lieberthal, Kenneth and Mikkal Herberg, *China's Search for Energy Security: Implications for U.S. Policy.* Seattle: National Bureau of Asian Research, Volume 17, Number 1, April 2006.

Lum, Thomas, Hannah Fischer, Julissa Gomez-Granger, and Anne Leland, *China's Foreign Aid Activities in Africa, Latin America, and Southeast Asia.* Congressional Research Service, the Library of Congress, R40361, February 25, 2009.

Maack, Jonathan N., "Scenario Analysis: A Tool for Task Managers," in *Social Analysis: Selected Tools and Techniques.* Washington, D.C.: World Bank, 2001, pp. 62–87.

Mao, Yufeng, "China's Interests and Strategy in the Middle East and the Arab World," in Joshua Eisenman, Eric Heginbotham, and Derek Mitchell, eds., *China and the Developing World: Beijing's Strategy for the Twenty-First Century.* Armonk, New York: M.E. Sharpe, 2007, pp. 113–32.

Marketos, Thrassy N., *China's Energy Geopolitics: The Shanghai Cooperation Organization and Central Asia.* New York: Routledge, 2009.

Massoud, Adla, "Oil May Fuel Sino-U.S. Conflict." *Al-Jazeera*, June 29, 2006.

Matheswaran, M., "Air and Aerospace Partnership" in Swaran Singh, ed., *China-Pakistan Strategic Cooperation: Indian Perspectives.* New Delhi, India: Manohar Publishers, 2007, pp. 195–208.

McNeal, Dewardic L., *China's Western Development Campaign.* Congressional Research Service, the Library of Congress, RL31104, September 12, 2001.

Medeiros, Evan S., *China's International Behavior; Activism, Opportunism, and Diversification.* Santa Monica, California: RAND, MG-350-AF, 2009.

Medeiros, Evan S., Keith Crane, Eric Heginbotham, Norman D. Levin, Julia F. Lowell, Angel Rabasa and Somi Seong, *Pacific Currents: The Responses of U.S. Allies and Security Partners in East Asia to China's Rise.* Santa Monica, California: RAND, MG-736-AF, 2008.

Military Advisory Board, *Powering America's Defense: Energy and the Risks to National Security.* Alexandria, VA: Center for Naval Analysis, May 2009.

Mix, Derek E., *The United States and Europe: Current Issues.* Congressional Research Service, Library of Congress, RS22163, March 16, 2009.

Moran, Daniel and James A. Russell, "Introduction: the militarization of energy security," in Daniel Moran and James A. Russell, eds., *Energy Security and Global Politics: the Militarization of Resource Management.* New York: Routledge, 2009, pp. 1–18.

Nanto, Dick K., James K. Jackson, Wayne M. Morrison, and Lawrence Kumins, *China and the CNOOC Bid for Unocal: Issues for Congress.* Congressional Research Service, the Library of Congress, RL33093, February 27, 2006.

National Academy of Engineering and National Research Council, *Personal Cars and China*. Washington, D.C.: National Academies Press, 2003.

National Petroleum Council, *Facing The Hard Truths About Energy: A Comprehensive View To 2030 Of Global Oil And Natural Gas (Draft Report),* Washington, D.C., July 18, 2007.

Ng Tze-Wei, "African Ties 'Threaten Influence of the West," *SCMP,* www.sinoptic.ch/embassy/presseschau/2006/20061211–1215.htm, December 15, 2006.

Niazi, Tarique, "The Ecology of Strategic Interests: China's Quest for Energy Security from the Indian Ocean and the South China Sea to the Caspian Sea Basin," China And Eurasia Forum Quarterly, Volume 4, No. 4 (2006), pp. 97–116.

Niazi, Tarique, "Gwadar: China's Naval Outpost on the Indian Ocean," *Jamestown Foundation*, February 28, 2005.

Nichol, Jim, *Armenia, Azerbaijan, and Georgia: Security Issues and Implications for U.S. Interests*. Congressional Research Service, the Library of Congress, RL30679, January 14, 2009.

Nichol, Jim, *Central Asia: Regional Developments and Implications for U.S. Interests*. Congressional Research Service, Library of Congress, RL33458, November 13, 2008.

Nichol, Jim, *Kazakhstan: Recent Developments and U.S. Interests*. Congressional Research Service, Library of Congress, 97–1058, June 20, 2008.

Nichol, Jim, *Turkmenistan's Political Succession: Developments and Implications*, Congressional Research Service, Library of Congress, RS22572, January 18, 2007.

Nichol, Jim, *Uzbekistan's Closure of the Airbase at Karshi-Khanabad: Context and Implications*, Congressional Research Service, Library of Congress, RS22295, October 7, 2005.

Nincic, Donna J., "Sea Lane Security and U.S. Maritime Trade: Chokepoints as Scarce Resources," in Sam J. Tangredi, ed., *Globalization and Maritime Power*. Honolulu: University Press of the Pacific, 2004, pp. 143–70.

Norlega, Roger F., *China's Influence in the Western Hemisphere* (Statement before the House Subcommittee on the Western Hemisphere). Washington, D.C.: U.S. Department of State, April 6, 2005.

Obiorah, Ndubisi, Darren Kew and Yushf Tanko, " 'Peaceful Rise' and Human Rights: China's Expanding Relations with Nigeria," in Robert I. Rotberg, ed., *China Into Africa: Trade, Aid and Influence*. Washington, D.C.: Brookings Institution Press, 2008, pp. 272–95.

Office of the Secretary of Defense, *Proliferation: Threat and Response 1996*. Washington, D.C.: U.S. Department of Defense, 1996.

Oliker, Olga and David A. Shlapak, *U.S. Interests in Central Asia: Policy Priorities and Military Roles*. Santa Monica, California: RAND, MG-338, 2005.

Oresman, Matthew, "Repaving the Silk Road: China's Emergence in Central Asia," in Joshua Eisenman, Eric Heginbotham, and Derek Mitchell, eds., *China and the Developing World: Beijing's Strategy for the Twenty-First Century*. Armonk, New York: M.E. Sharpe, 2007, pp. 60–83.

Organization for Economic Co-Operation and Development, *The Export Credits Arrangement*, 1978–2008.

Palacios, Luisa, "Latin America as China's Energy Supplier," in Riordan Roett and Guadalupe Paz, eds., *China's Expansion into the Western Hemisphere: Implications for Latin America and the United States.* Washington, D.C.: Brookings Institution Press, 2008, pp. 170–89.

Pan, Esther, *China, Africa and Oil.* New York: Council on Foreign Relations, January 12, 2006.

Pehrson, Christopher J., *String of Pearls: Meeting the Challenge of China's Rising Power across the Asian Littoral.* Carlisle, PA: Strategic Studies Institute, July 2006.

Peters, John E., James Dickens, Derek Eaton, C. Christine Fair, Nina Hachigian, Theodore W. Karasik, Rollie Lal, Rachel M. Swanger, Gregory F. Treverton, and Charles W. Wolf, *War and Escalation in South Asia.* Santa Monica, California: RAND, MG-367–1, 2006.

Pham, J. Peter, "China's African Strategy And Its Implications for U.S. Interests," *American Foreign Policy Interests,* Issue 28, 2006, pp. 239–53.

Prabhakar, W. Lawrence, "The Maritime Convergence" in Swaran Singh, ed., *China-Pakistan Strategic Cooperation: Indian Perspectives.* New Delhi, India: Manohar Publishers, 2007, pp. 231–47.

Puska, Susan, "Military Backs China's African Adventure," *Asia Times Online*, June 8, 2007.

Puska, Susan, "Resources, Security and Influence: The Role of the Military in China's Security Strategy," *China Brief*, Jamestown Foundation, Volume VII, Issue 11, May 30, 2007.

Rahim, Saad, "Regional Responses and Strategic Responses," in Daniel Moran and James A. Russell, eds., *Energy Security and Global Politics: The Militarization of Resource Management.* New York: Routledge, 2009, pp. 95–111.

Ratliff, William, "An Assessment of China's Deepening Ties to Latin America," *China Brief*, Jamestown Foundation, Volume VIII, Issue 11, May 22, 2008.

Rieber, Steven, "How Statistical Models Can Help Intelligence Analysts," Georgia State University, 2004.

Roett, Riordan and Guadalupe Paz, "Introduction: Assessing the Implications of China's Growing Presence in the Western Hemisphere," in Riordan Roett and Guadalupe Paz, eds., *China's Expansion into the Western Hemisphere: Implications for Latin America and the United States.* Washington, D.C.: Brookings Institution Press, 2008, pp. 1–23.

Root, Hilton, L., *Alliance Curse: How America Lost the Third World.* Washington, D.C.: The Brookings Institution, May 1, 2008.

Root, Hilton, L., *Capital and Collusion: The Political Logic of Global Economic Development.* Princeton, NJ: Princeton University Press, November 21, 2005.

Rose, Thom, "Saudi Arabia Cuts Oil Sales To U.S., Ups China," *Energy Bulletin,* www.energybulletin.net/newswire.php?id=2348, September 15, 2004.

Rotberg, Robert I., "China's Quest for Resources, Opportunities, and Influence in Africa" in Robert I. Rotberg, ed., *China into Africa: Trade, Aid and Influence.* Washington, D.C.: Brookings Institution Press, 2008, pp. 1–20.

Sakhuja, Vijay, "Naval Cooperation" in Swaran Singh, ed., *China-Pakistan Strategic Cooperation: Indian Perspectives*. New Delhi, India: Manohar Publishers, 2007, pp. 209–30.

Sandalow, David, *Ending Oil Dependence*. Washington, D.C.: The Brookings Institution, January 22, 2007.

Santoli, Al, *The Panama Canal in Transition: Threats to U.S. Security and China's Growing Role in Latin America*. Washington, D.C.: American Foreign Policy Council, June 23, 1999.

Saunders, Philip C., *China's Global Activism: Strategy, Drivers, and Tools*, Washington, D.C.: Institute for National Strategic Studies, National Defense University, 2006.

Schwartz, Peter, *The Art of the Long View: Planning for the Future in an Uncertain World*, New York: Currency Doubleday, 1996.

Sevastopulo, Demetri, "US Concerns over China Weapons in Iraq," *Financial Times*, July 7, 2007.

Shambaugh, David, Eberhard Sandschneider, and Zhou Hong, eds., *China-Europe Relations: Perceptions, Policies and Prospects*. New York, New York: Routledge, 2008.

Shinn, David H., "Military and Security Relations: China, Africa, and the Rest of the World" in Robert I. Rotberg, ed., *China into Africa: Trade, Aid, and Influence*. Washington, D.C.: Brookings Institution Press, 2008, pp. 155–96.

Sidhu, Waheguru Pal Singh and Jing-dong Yuan, *China and India: Cooperation or Conflict?* London: Lynne Rienner Publishers, Inc., 2003.

Singh, Swaran, "Introduction" in Swaran Singh, ed., *China-Pakistan Strategic Cooperation: Indian Perspectives*. New Delhi, India: Manohar Publishers, 2007, pp. 17–34.

Sinton, Jonathan, Rachel E. Stern, Nathaniel T. Aden and Mark D. Levine, *Evaluation of China's Energy Strategy Options*, Lawrence Berkeley National Laboratory, Berkeley, California, May 2005.

Small, Andrew, "The US Factor," in Stanley Crossick and Etienne Reuter, eds., *China-EU: A Common Future*. Singapore: World Scientific Publishing, 2007, pp. 73–80.

Solingen, Etel, "From 'Threat to Opportunity'? ASEAN, China, and Triangulation," in Evelyn Goh and Sheldon W. Simon, eds., *China, United States, and Southeast Asia: Contending Perspectives on Politics, Security, and Economics*. New York: Routledge, 2008, pp. 17–37.

Spencer, Richard, "China to Take More Active Role in Africa," www.telegraph.co.uk/news/main.jhtml?xml=/news/2007/05/19/wchina19.xml, May 19, 2007.

Stanzel, Volker, "The EU and China in the Global System," in David Shambaugh, Eberhard Sandschneider and Zhou Hong, eds., *China-Europe Relations: Perceptions, Policies and Prospects*. New York: Routledge, 2008, pp. 251–69.

Su, Sherry, "China Oil Firms Building Presence in Myanmar," *Dow Jones Energy Service*, January 16, 2007.

Sullivan, Mark P., *Venezuela: Political Conditions and U.S. Policy*. Congressional Research Service, Library of Congress, RL32488, August 1, 2008.

Sutter, Robert G., *Chinese Foreign Relations: Power and Policy since the Cold War*. Lanham, Maryland, Rowman & Littlefield Publishers, 2008.

Sutter, Robert G., *Chinese Policy Priorities and Their Implications for the United States*. Lanham, Maryland: Rowman & Littlefield Publishers, 2000.

Sutter, Robert G., *China's Rise: Implications for U.S. Leadership in Asia*. Washington, D.C.: East-West Center, 2006.

Sutter, Robert, "China's Rise, Southeast Asia, and the United States: Is a China-Centered Order Marginalizing the United States," in Evelyn Goh and Sheldon W. Simon, eds., *China, United States, and Southeast Asia: Contending Perspectives on Politics, Security, and Economics*. New York: Routledge, 2008, pp. 91–106.

Taylor, Ian, *China's New Role in Africa*. London: Lynne Rienner Publishers, Inc., 2009.

Teng, Chung-Chian, "Hegemony or Partnership: China's Strategy and Diplomacy Toward Latin America," in Joshua Eisenman, Eric Heginbotham, and Derek Mitchell, eds., *China and the Developing World: Beijing's Strategy for the Twenty-First Century*. Armonk, New York: M.E. Sharpe, 2007, pp. 84–112.

Tetlock, Philip E., *Expert Political Judgment: How Good Is It? How Can We Know?* New Jersey: Princeton University Press, 2005.

Tiagi, Raaj and Lu Zhou, *Canada's Economic Relations with China*. Canada: Fraser Institute, February 2009.

Tokatlian, Juan Gabriel, "A View from Latin America," in Riordan Roett and Guadalupe Paz, eds., *China's Expansion into the Western Hemisphere: Implications for Latin America and the United States*. Washington, D.C.: Brookings Institution Press, 2008, pp. 59–89.

Tu, Jianjun, "China's Strategic Petroleum Reserves in Sino-Venezuela Relations," *China Brief*, Jamestown Foundation, Volume VIII, Issue 19, October 7, 2008.

USA Today, "Chinese Economy Slows to Still Sizzling 11.5% Growth," September 25, 2007.

U.S. Agency For International Development, *U.S. Overseas Loans and Grants: Obligations and Loan Authorizations July 1, 1945–September 30, 2005*, PN-ADH-500, pp. 1–231, 2005.

U.S. Central Intelligence Agency, *Chiefs of State and Cabinet Members of Foreign Governments: A Directory, DI-CS-2008–9*, 2008.

U.S. Congress, Congressional Budget Office, *China's Growing Demand for Oil and Its Impact on U.S. Petroleum Markets* (CBO), April 2006

U.S. Congress, Subcommittee on Africa, Global Human Rights and International Relations of the Committee on International Relations, *China's Influence in Africa*, Hearing on July 28, 2005, 109th Congress, Session 1, 2005, pp. 1–77.

U.S. Congress, Subcommittee on the Middle East and South Asia, *U.S. Relations with Saudi Arabia: Oil, Anxiety, and Ambivalence*, Hearing on September 18, 2007, 110th Congress, Session 1, 2007, pp. 1–42.

U.S. Congress, Subcommittee on the Western Hemisphere, *China's Influence on the Western Hemisphere*, Hearing on April 6, 2005, 109th Congress, Session 1, 2005, pp. 1–78.

U.S. Congress, U.S.-China Economic and Security Review Commission, *China's Energy Needs and Strategies*, Hearing on October 30, 2003, 108th Congress, Session 1, 2003, pp. 1–135.

U.S. Department of Energy, *Energy Policy Act 2005-Section 1837: National Security Review of International Energy Requirements*. U.S. Department of Energy, February 2006.

U.S. Department of Energy and U.S. Environmental Protection Agency, *Fuel Economy Guide 2009*.

U.S. Department of State, Background Notes: Angola, www.state.gov, December 2008.

U.S. Department of State, Background Notes: Canada, www.state.gov, November 2008.

U.S. Department of State, Background Notes: India, www.state.gov, January 2009.

U.S. Department of State, Background Notes: Iraq, www.state.gov, February 2008.

U.S. Department of State, Background Notes: Nigeria, www.state.gov, December 2008.

U.S. Department of State, Background Notes: Pakistan, www.state.gov, March 2009.

U.S. Department of State, Background Notes: Saudi Arabia, www.state.gov, February 2008.

U.S. Department of State, Background Notes: Sudan, www.state.gov, January 2009.

U.S. Department of State, Background Notes: Venezuela, www.state.gov, January 2009.

Energy Information Administration, U.S. Department of Energy, Country Analysis Briefs: Mexico, www.eia.doe.gov, March 2009.

U.S. Environmental Protection Agency, *Clean Automotive Technology: Innovation That Works*, EPA-420-F-04–024, March 2004.

"U.S. Says Chinese Missiles Smuggled through Iran into Iraq," www.defencetalk.com/news/publish/index.php, July 23, 2007.

U.S. Senate, Committee on Foreign Relations, *The Emergence of China throughout Asia: Security and Economic Consequences for the United States*, Hearing on June 7, 2005, 109th Congress, Session 1, 2005, pp. 1–62.

U.S. Senate, Committee on Foreign Relations, *Energy Trends in China and India: Implications for the United States*, Hearing on July 26, 2005, 109th Congress, Session 1, 2005, pp. 1–57.

"Venezuela's Chavez Arrives in China; Energy a 'Strategic Line'," *Caracas Agencia Bolivariana de Noticias*, August 22, 2006.

Verma, Virendra Sahai, "The Karakoram Highway" in Swaran Singh, ed., *China-Pakistan Strategic Cooperation: Indian Perspectives*. New Delhi, India: Manohar Publishers, 2007, pp. 331–44.

Watkins, Eric, "China 'Stocking Up' on World Oil, Analysts Say," *Oil and Gas Journal*, March 16, 2009, Volume 107.11, Page 3.

Watson, Cynthia, "The PLA in Latin America," *China Brief*, Jamestown Foundation, Volume VII, Issue 20, October 31, 2007.

Weintraub, Sidney, *The Energy Situation in the Western Hemisphere*. Washington, D.C.: CSIS Press, July 2007.

Xiang, Lanxin, "An Alternative Chinese View," in Riordan Roett and Guadalupe Paz, eds., *China's Expansion into the Western Hemisphere: Implications for Latin America and the United States*. Washington, D.C.: Brookings Institution Press, 2008, pp. 44–58.

Xu, Xiaojie, "China and the Middle East: Cross-Investment in the Energy Sector," *Middle East Policy*, Vol. VII, No. 3, June 2000, pp. 122–36.

Yamaguchi, Nancy, David Fridley, and Ke Xiaoming, *Improving Transport Fuel Quality in China: Implications for the Refining Sector*, Lawrence Berkeley National Laboratory, August 2002.

Yardley, Jim, "China's Leader Signs Oil Deals with Africans," *New York Times*, May 1, 2006.

Yergin, Daniel, "Ensuring Energy Security," *Foreign Affairs*, March/April 2006, pp. 69–82.

Yergin, Daniel, "Over a Barrel," *Fortune*, Volume 151, Issue 10, May 16, 2005, p. 114.

Yergin, Daniel, "Riding the Tiger: The Global Impact of China's Energy Quandary" (International), *Energy*, Volume 29, Issue 2, March 22, 2004, pp. 34–38.

Yergin, Daniel, *The Prize: The Epic Quest for Oil, Money and Power*. New York: Free Press, 1992.

Yetiv, Steve A. and Chunlong Lu, "China, Global Energy and the Middle East," *The Middle East Journal*, Volume 61, No. 2, Spring 2007, pp. 199–218.

Zambelis, Chris and Brandon Gentry, "China through Arab Eyes: American Influence in the Middle East," *Parameters*, Spring 2008, pp. 60–72.

Zha Daojiong, "China's Energy Security and Its International Relations," *China and Eurasia Forum Quarterly*, Volume 3, No. 3, November 2005, pp. 39–54.

Zhang, Xiaodong, "China's Interests in the Middle East: Present and Future," *Middle East Policy*, Vol. 6, No. 3, February 1999, pp. 150–59.

Zheng, Denise E., "China's Use of Soft Power in the Developing World: Strategic Intentions and Implications for the United States," in Carola McGiffert, ed., *Chinese Soft Power and Its Implications for the United States*. Washington, D.C.: CSIS Press, March 2009, pp. 1–9.

Zhou Fengqi, "Strategic Considerations for Development of Renewable Energy in China," Zhongguo Kexueyuan Yuankan, Vol 21 No 4, July 15, 2006, pp. 287–94.

Zulay Rosas, "Venezuela: President Chavez, Officials Outline Objectives of China Visit," *Caracas Venezolana de Television*, August 23, 2006.

Zweig, David and Bi Jianhai, "China's Global Hunt for Energy," *Foreign Affairs*, September/October 2005, pp. 25–38.

Index

About the Author

George Eberling is a 1985 graduate of The Citadel. He completed his Ph.D. in political science from Claremont Graduate University in May 2010. *Chinese Oil Energy Futures and Their Implications for the United States* is his first publication.

Lightning Source UK Ltd.
Milton Keynes UK
UKOW051333060213

205922UK00001B/49/P

9 780739 165683